Elements of
ANIMAL BEHAVIOUR

Elements of ANIMAL BEHAVIOUR

M. Nithya Devi

ANMOL PUBLICATIONS PVT. LTD.
NEW DELHI-110 002 (INDIA)

ANMOL PUBLICATIONS PVT. LTD.
Regd. Office: 4360/4, Ansari Road, Daryaganj,
New Delhi-110002 (India)
Tel.: 23278000, 23261597, 23286875, 23255577
Fax: 91-11-23280289
Email: anmolpub@gmail.com
Visit us at: www.anmolpublications.com

Branch Office: No. 1015, Ist Main Road, BSK IIIrd Stage
IIIrd Phase, IIIrd Block, Bengaluru-560 085 (India)
Tel.: 080-41723429 • Fax: 080-26723604
Email: anmolpublicationsbangalore@gmail.com

Elements of Animal Behaviour

First Edition, 2011

PRINTED IN INDIA

Printed at Mehra Offset Press, Delhi.

Contents

Preface

Animal behaviour is the scientific study of everything animals do, whether the animals are single-celled organisms, invertebrates, fish, amphibians, reptiles, birds, or mammals. It involves investigating the relationship of animals to their physical environment as well as to other organisms, and includes such topics as how animals find and defend resources, avoid predators, choose mates and reproduce, and care for their young.

People who study animal behaviour are concerned with understanding the causes, functions, development, and evolution of behaviour. The causes of behaviour include both the external stimuli that affect behaviour, and the internal hormonal and neural mechanisms that control behaviour. The functions of behaviour include its immediate effects on animals and its adaptive value in helping animals to survive or reproduce successfully in a particular environment. The development of behaviour pertains to the ways in which behaviour changes over the lifetime of an animal, and how these changes are affected by both genes and experience. The evolution of behaviour relates to the origins of behaviour patterns and how these change over generations.

Studies on animal behaviour are vital for several reasons most importantly in connection with conservation, livestock and wildlife management, animal welfare, sustainable use of animal and human resources, and ultimately for better understanding of human behaviour.

Author

1

Ecology of Animal Behaviour

The predictable behavioural Programmes are inherited by animals through their parents and portions of the Programmes are open to natural selection and modification. Thus, these behaviours are phylogenetic adaptations that have an evolutionary history. This often leads to a comparative approach and has led researchers to search for the biological basis of human behaviour by comparing our activities to those of our close relative (other primates; especially chimps). There are two schools of thought as to how animals acquire their behaviour patterns.

Some hold to the view that animals, including humans, learn all their behaviour during the course of ontogenetic development. Ducks, for example, learn how to quack like a duck and don't honk like a goose because they hear their parents while within the egg). Experiments have since shown that these behaviours are built-in and not learned.

Very complex behaviour patterns can be passed on through the genes. A spider's web, for example, is built perfectly the first time a spider attempts construction, despite the fact that they may have no prior experience with webs. Female spiders construct egg sacs in the fall, and then die. The spiderlings emerge in the spring, and never had experience with an orb web before building their own. Ethology differs from the study of Animal Behaviour, in that animals behaviourists generally are interested in learned behaviours while ethologists concentrate on innate behaviours. Also, animals behaviourists tend to be trained in psychology, while ethologists are zoologists. Animal behaviourists tend to work with "bright" animals, such as rats; putting them through trials with mazes and Skinner boxes. While the study of learned behaviour is both important and immediately applicable to human psychology, these behaviours cannot have an evolutionary basis other than the neural capacity to learn.

TERRITORIAL BEHAVIOUR

Territorial behaviour has the potential to reduce the number of individuals that a given area can support; thus the ecological and social circumstances that promote territoriality will affect population persistence. Eadie, Sherman, and Semel highlight a counterintuitive consequence of territoriality in cavity-nesting birds.

When artificial next sites are provided to increase population size, species lacking territorial behaviour suffer from high rates of conspecific brood parasitism, which lowers reproductive success, whereas territorial defence of nest-boxes prevents parasitism, enabling the population to benefit from this management procedure. Baseline behavioural and ecological data collected in the course of field studies have the potential to predict a population's response to habitat disturbance.

For example, frequency of intraspecific interactions may affect spread of disease within populations. Harcourt attempts to determine which ecological and behavioural variables successfully predict primate species' ability to survive in selectively logged forest. Similarly, it has been argued that information about a species' behaviour under different ecological and social conditions is the key to success of reintroductions.

Indeed, Curio suggests that a better understanding of antipredator behaviour and its development would lead to greater survival and reproduction of captive birds released into the wild. Finally, many problems in *ex situ* conservation have been solved or ameliorated using knowledge of activity patterns, diet, and social behaviour. Wielebnowski systematically reviews important areas of behavioural research that are crucial to captive propagation of mammals, including space requirements, social organization, parental care, and sexual behaviour.

ANTIPREDATOR BEHAVIOUR

Predation is an important source of mortality in the wild. Many species of mammals and flightless birds have gone extinct on islands as a result of introduced domestic cats, rats, and mongooses. These endemics lost certain aspects of predator avoidance behaviour as they evolved on islands without terrestrial predators; preliminary evidence suggests that some continental prey species are also losing their ability to respond to cues of predators in places where large carnivores have been extirpated.

Understanding antipredator behaviour may allow us to make predictions about the fate of populations subjected to introduced predators or conservation manipulation. Controversial data suggest that black rhinoceros Diceros bicornis calves suffer higher mortality in the presence of large carnivores if their mothers are dehomed as part of a conservation procedure, presumably as a result of mothers' inability to fend off predatory attack. Prior knowledge of maternal antipredator behaviour and the function of horns in females would have spoken against dehorning. Since grouping confers advantages on individuals through early detection of predators, warning conspecifics of danger, reducing vigilance time, and diluting predation risk variation in patterns of grouping make species and age-sex classes differentially vulnerable to hunting by humans.

FitzGibbon discusses how antipredator strategies affect the probability of being selected by subsistence hunters. She notes that individuals respond to predation risk by reducing time spent in other activities such as feeding but the extent to which human hunting pressure indirectly affects reproductive rates in this way remains unexplored. Whether antipredator behaviour is principally innate or is modified by experience is critical to captive release Programmes and some attempts have been made to train captive-bred individuals to recognize and avoid predators. Although we know that predator recognition is innate in some species such as motmots *Eumomota superciliosa* but must be learned in others such as vervet monkeys *Cercopithecus aethiops* and Belding's ground squirrels *Spertnophilus beldingi* the ecological circumstances under which different sorts of predator template development occurs are still poorly understood. Innate recognition may facilitate rapid antipredator responses, but it may limit our ability to transfer local stocks between areas with different predators. Learned predator recognition, in contrast, may result in high mortality initially, but it does allow reintroduced prey to respond to predators or free-living prey to cope with exotic or reintroduced predators.

FORAGING BEHAVIOUR

Foraging behaviour affects population persistence in the general sense that carnivores are at greater risk of extinction than herbivores, and that, within a dietary class, specialist foragers are more extinction prone than generalists. Thus knowledge of the food requirements of rare species is useful. At a general level, habitat preferences contingent on feeding requirements will influence species' responses to habitat disturbance and fragmentation. More specifically, foraging theory has shown how patterns of interference between competitors and depletion of feeding patches predict the distribution of individuals and hence the number that a site can hold. Moreover, models have been developed to predict the distribution of animals in patches of differing profitability when individuals are equally competitive or when some dominate others at resources. In theory, these tools might allow conservation biologists to predict consequences of habitat fragmentation if factors such as food abundance, food distribution, and individuals' competitive abilities were known. Monaghan has shown how patterns of feeding in seabirds such as foraging trip length and number of dives can be used to monitor marine fish populations.

DEMOGRAPHIC VARIABLES

Behavioural ecological field studies usually have records

on the age and sex of individuals, so it is relatively easy to construct the age structure of the population and the sex ratios of juvenile and adult segments. Estimates of Ne are greatly influenced by population demographics such as these.Creel uses long-term demographic records from social carnivore species to investigate the demographic factors affecting Ne. Because behavioural ecologists are interested in the consequences of traits on individual reproduction and survival, studies usually strive to collect detailed records on individuals' age at first reproduction, interbirth intervals, litter sizes, offspring sex ratios, and longevity, although information on age-specific mortality is more difficult to collect. Life history variables are necessary for the construction of PVAs used to calculate both demographic and genetic MVPs.

Although influential and early PVAs used data collected by field biologists to make predictions about population persistence they were based on relatively crude empirical data. In the context of politically charged controversies where models are scrutinized closely, resulting recommendations may be called into question.

Detailed information on individual life histories can remedy this because they provide an exact measure of variance, which greatly increases the resolution of PVA and Ne calculations.Durant highlights the importance of life history variables in affecting population persistence using three mammal species with different social structures. Her models show that skewed sex ratios are a powerful indicator of extinction and that extending adult survival is effective in increasing population growth rates.

COMPETITION BETWEEN MALES

Certain behavioural and morphological traits are solely used in competition for access to members of the opposite sex. Contests between males have led to the evolution of weaponry such as beetle and deer antlers, to infanticidal behaviour, in which males kill offspring fathered by competitors, and to mate guarding. In some species, high costs of fighting have selected for polymorphic fighting tactics in which some males compete severely over females, whereas others avoid confrontation and mate surreptitiously. The outcome of contests between males will affect who breeds and hence variance in male reproductive success, which affects Ne. Another example of how variance in male reproductive success affects genetic components of populations comes from bison *Bison bison* in the Badlands National Park.

There, one lineage is currently being lost because males of that lineage are more timid and less able to defend females than those of the more successful line. Knowledge of patterns of mating and the cues that individuals use to select mates can also help identify the potential for hybridization between species; because one of the goals of conservation is species integrity and persistence, hybridization is normally viewed with alarm. Greene, Umbanhowar, Mangel, and Caro show that in infanticidal species where males kill offspring fathered by other males, population growth rates are reduced, and it may be difficult for such species to withstand exploitation. In theory, Ne will be lowered in species in which infanticide or mate guarding is prevalent.

Different fighting tactics are usually associated with different mortality schedules; thus the frequency with which males display different behavioural and morphological tactics will affect population growth rate and Ne. Rubenstein draws attention to the fact that those Atlantic salmon *Salmo salar* migrating to the sea are more likely to be caught by fishermen than conspecifics that remain behind and mature in their natal streams and that this differential offtake selects for philopatric phenotypes. In a number of reef-dwelling fishes, intersexual competition for mating opportunities has led to the evolution of protandry, in which individuals change from male to female, or protygyny, in which they change from female to male.

Sex change may be precipitated by reaching a certain age or size or by social circumstances when it pays individuals to become the currently rarer sex. Vincent and Sadovy argue that exploitation of larger fish for the tropical fish trade may remove males of certain species, for example, from a population. Thus the speed with which females are able to change to being male will be crucial in determining a population's response to exploitation.

MATE CHOICE BY FEMALES

Among females, mate choice is increasingly seen as a means to enhance offspring fitness through choice of males with heritable, condition-enhancing traits such as resistance to parasites. In the wild, loss of males from small populations may theoretically limit the range of males from which to choose to the extent of slowing the rate of population increase. In captive populations, some degree of female choice may be advisable because, as managers, we are poor at picking out healthy stock, at identifying traits that enhance fitness, and we have little to no knowledge of trait heritability.

Grahn, Langefors, and von Schantz argue that failing to allow for mate choice in captive salmon stocks has resulted in disease-induced mortality. Mate choice has possible implications for captive breeding of endangered birds, too. Zebra finches *Poephila guttata* show preferences for mating with individuals wearing particular colour bands and will adjust the amount of parental investment and sex ratio of their offspring according to the colour of their partner's leg bands. This may upset population projections of *in situ* breeding Programmes.

Recently, considerable attention has focused on the relationship between mate choice and the degree of symmetry in male phenotypic traits. This is because symmetrical traits are thought to reflect developmental stability and hence the ability of the genotype to control development under a range of environmental conditions. There is evidence that symmetrical ornaments correlate with fitness and that females choose males on the basis of ornament symmetry. Although behavioural ecologists are interested in trait asymmetry as a marker of genetic quality, asymmetry also reflects the extent of environmental insult during development and, measured repeatedly, can provide an index of a population's response to environmental change over time. Furthermore, asymmetric leg bands placed on captive male birds may modify female choice and could result in lower reproductive output in endangered species.

MATING SYSTEMS

The distribution of resources crucial to females, and hence the economic defendability of females by males interact with patterns of parental care to produce the mating system. Because Ne is lowered by increasing skew in the ratio of breeding females to breeding males, whether a population is polygynous or monogamous, or more specifically, the extent of extrapair paternity, therefore has a strong bearing on loss of genetic diversity. In addition, Ne is lowered by increasing variance in family size, which is marked in reproductively suppressed species. Creel systematically explores aspects of the mating system that affect Ne using demographic data on wild carnivores. He reports that unequal reproductive success causes the greatest reduction in Ne, with skewed sex ratios and variation in population size being less important. Differences in mating system will also affect a population's ability to sustain commercial exploitation or harvesting. As an example, population growth rates of monoganıous species are likely to be less resilient to removal of individuals than are those of polygynous species because in monogamous species there are fewer males available for remating and parental care.

PARENTAL CARE AND SEX ALLOCATION

Theory underlying the evolution of parental care, including intraspecific variation in care in relation to costs and benefits, parent-offspring conflict, and differential investment in sons and daughters, is well understood.The sex that cares for offspring and the period of time over which care occurs will affect a species' ability to withstand offtake of a given magnitude and selectivity. Vincent and Sadovy suggest that loss of parental males could reduce the reproductive output of an exploited population in taxa such as seahorses in which male care limits female reproductive rate.

Adaptive alterations in offspring production can affect conservation agendas in at least two ways. First, in populations that are selectively exploited, parents may bias offspring production in favour of the rarer sex. For example, Creel and Creel have shown that in lions birth sex ratios are male biased in populations hunted for male trophies, and this has ramifications for sustainable offtake.

Rubenstein shows how failure to account for sex ratio adjustment in onagers *Equus hemionus* jeopardized a reintroduction Programme. Second, bird recovery Programmes make use of a female's disposition to lay replacement clutches if her first brood is removed, and they also capitalize on hatching asynchrony to cross-foster the first hatchling to a conspecific or heterospecific. Both management strategies can improve reproductive output of endangered species. In addition, knowledge of siblicide in birds is useful in preventing reproductive loss when trying to breed endangered birds.

DISPERSAL AND INBREEDING AVOIDANCE

In theory, one might expect the rate and distance over which animals disperse and its associated costs to have greater ramifications for conservation strategies than any other behavioural ecological variable. Rates of dispersal between habitat patches affect population persistence gene flow between subgroups influences heterozygosity and genetic drift, and the length and direction dispersers travel and their habitat preferences during movement speak to reserve design and connectivity.

Although dispersal is often studied in animal and plant populations high-quality data are, unfortunately, difficult to obtain; for example, maximal dispersal distances are usually underestimated in behavioural studies. Outside of PVAs, there have been few attempts to use empirical data on dispersal in conservation theory or practice. To remedy this Van Vuren outlines the way in which dispersal distance, dispersal direction, and survival of dispersers might influence design of reserves and managing areas between them.

As populations become increasingly isolated, the threat of inbreeding increases. Knowledge of the extent of philopatry and strength of inbreeding avoidance mechanisms across taxa can identify which species are most likely to be subject to inbreeding depression.

This would then prioritize which species or populations require translocations and on which species noninvasive genetic screening might be carried out, as advocated by Waldman and Tocher. In addition, the extent to which different species tolerate inbreeding in the wild informs captive Programmes about the relative importance of avoiding pairing relatives in zoo breeding Programmes.

COOPERATION AND HELPING

Many examples of cooperative breeding in birds and mammals are a consequence of habitat saturation, a situation increasingly likely to occur as habitats become isolated and smaller in area through fragmentation. Populations of cooperative breeders are likely to have lower population growth rates than populations where all individuals pair up. For instance, Seychelles warblers *Acrocephalus sechellensis* transferred from densely populated Cousin Island to empty Aride Island exhibited faster rates of reproduction than their founding population.

On the other hand, the presence of helpers or floating nonresident individuals can act as a buffer to small decreases in population size, but the dynamics of these conflicting effects are unexplored. Knowledge of the circumstances under which cooperative breeding will appear in different species may therefore inform intervention strategies, while reproductive benefits that parents receive from offspring can help zoo breeding Programmes set up appropriate social groups.

METHODS IN BEHAVIOURAL ECOLOGY

Behavioural ecology focuses on the adaptive significance of variation in individual behaviour, morphology, and physiology which has forced scientists to learn to recognize individual animals.

Consequently, all the individuals within a subpopulation are often known to researchers fortuitously providing an accurate estimate of local population size. Individually recognizing all members of a population is a far more accurate method of estimating population size than using classical ecological techniques such as mark-recapture methods or stratified sampling.

In addition, long-term field studies have unusually accurate figures of changes in population size in turn, this may influence effective population size. McGregor and Peake review methods used to recognize animals individually and additionally show that populations can be monitored through individual call recognition rather than observing animals themselves.

MOLECULAR TECHNIQUES

In studies of known individuals in the wild, data are regularly collected on paternity, and sometimes maternity, using genetic markers. These data generate information on the extent to which particular individuals contribute to the next generation and hence to N. Knowledge of the number of lineages, or breeding groups, for a given Ne will alter estimates of the time frame over which allelic diversity is maintained. Allozymes, DNA fingerprinting, mitochondrial DNA (mtDNA), minisatellite, and microsatellite techniques shed light on population differentiation and genetic variability.

These methods have identified subpopulations of humpback whales *Megaptera novaeangliae)* migration routes and extent of philopatry in marine turtles, and the degree of hybridization between canid species information almost impossible to obtain from observation alone. In addition, analyses of mtDNA sequences determine phylogenetic relationships between species, which help identify evolutionarily significant units (ESUs) for conservation. For instance, genetic analyses of orangutan *Pongo pygmaeus* populations indicate sufficient divergence between Sumatran and Bornean populations to put a halt to breeding individuals from the two regions.

Although within-and between-species molecular comparisons are normally outside the purview of behavioural ecology, analyses that attempt to uncover associations between behavioural and morphological variables without the confounding effects of shared ancestry nonetheless require phylogenetic trees derived using molecular techniques.

HUMAN BEHAVIOURAL ECOLOGY

Because anthropogenic factors are the greatest challenge to biodiversity, many conservation biologists would argue that conservation solutions will eventually emerge from understanding human behaviour. Patterns of human exploitation are better understood now than they were 15 years ago. The myth of the "noble savage" living in harmony with the environment has been exploded.

In a detailed exploration of this issue, Alvard uses empirical data to show that Piro hunters in Peru pursue an optimal foraging strategy rather than any conservation agenda. As a second illustration, the tragedy of the commons, which describes individuals overexploiting a common property, is better understood in terms of a problem of unrestricted access rather than a problem of ownership. This focuses attention on the best ways to police resources rather than on who owns them.

These findings are informative in that they show which conservation strategies are likely to succeed or fail. In spite of these advances, the rapid growth of evolutionary psychology, a subdiscipline in which contemporary human behaviour is interpreted in the light of past selection pressures, has so far had little to say about patterns of current exploitation. Similarly, economics does not consider the underlying biological motivation in decision making.

Nevertheless, it seems likely that we will be able to understand human resource use better by paying closer attention to evolved behavioural dispositions. Specifically, they draw attention to differences in attitudes towards conservation shown by different segments of society as predicted by evolutionary theory. Certainly, the two pivotal problems in conservation biology, human population growth and resource accrual, are central to human behavioural ecology's focus on the social and ecological factors that affect reproductive success in different cultures. Demographic records and information on dispersal are obviously of great importance, while population growth rate, a population's response to exploitation, and Ne are influenced by many factors in behavioural ecology.

There are other bridges between these disciplines that have not been included, however, and many others that doubtless will appear as our knowledge of behavioural adaptations improves.

Moreover, I think it fair to say that any behavioural ecologist with a conscience must be interested in the disappearance of biological diversity and, with a little thought, should be able to recognize the conservation significance of his or her research and make some contribution to solving the crisis at hand.

2

Development of Physical Structure

The great difference between the two branches of physiology is this—that while adult physiology is dealing with stability, developmental physiology is dealing with change. In the former case, the general characters of the organism, in spite of alterations in detail, remain the same over long periods of time; and one of the most prominent features of the processes of adult life is that, as a whole, they constitute a self-regulating system, in which any excess in one direction automatically brings into play forces acting in the opposite direction. But in developmental physiology, the organism is always passing from one phase into another, and even if some of the phases—like the tadpole stage of the frog—are of comparatively long duration, yet processes are always at work which at length will upset the self-regulative power of the system, and will make it change into something quite different.

Besides these processes of normal development, however, others of the same general type may occur as the result of special circumstances. The facts of regeneration, for instance, come under this head; and, as a matter of fact, some analysis of regeneration will be found to give the best starting-point for a discussion of developmental physiology in general. Regeneration is not a very familiar fact to most of us, since it is present to a very slight degree in ourselves and in the animals we know best.

But worms or polyps, or indeed most of the lower animals—if they could think—would find regeneration the natural, normal state of affairs, and the absence of regeneration a sad and abnormal failure. In a sense they would be right.

Regeneration does at least seem to be a fundamental and original property of life, lost late and for special reasons. If you throw crumpled paper into a freshwater ditch, or set traps consisting of a bottle with boiled earth-worm for bait, you will probably succeed in catching large numbers of various kinds of freshwater Planarians—little leaf-shaped creatures of carnivorous habits and gliding movements. Choose out the brown and black species, which are the hardiest.

Transfer one to moist blotting-paper, and with a sharp knife cut it clean across. The experiment sounds cruel; but in reality you will have artificially encouraged the multiplication of the species. Each of the bits, replaced in water, will regenerate what is lost and so become transformed into a complete worm. The same would have happened if you had cut the animal into half a dozen cross-pieces instead of two, and equally whether you cut it down the middle or across. If you persisted, you would discover two main facts in the course of your experiments—that any piece of any shape, provided it was above a certain quite small size, is capable of regenerating into a whole animal; and that this happens equally whether the regenerating process are fed or no. The same would have been true if instead of a planarian you had cut up the little polyp Hydra, and almost the same with the smaller Annelid worms such as Lumbriculus. The same also holds for the microscopic single-celled Protozoa, although here the operations are much more difficult owing to the animals' small size.

It is also true for the very early stages of development in many kinds of animal. In a newt either of the two first cells produced in segmentation may, if artificially separated, give rise to a whole embryo; the same holds in many animals for any one of the first four cells, or for most fragments of the late segmentation and blastula stages. Thus regeneration of the extreme type—what we may call complete regenerative capacity—is found in the simplest types of animal and the earliest stages of development. It must therefore be thought of, not as a special property developed to meet the rare contingency of losing a limb or being bitten in two, but as something natural to living things, and present unless circumstances forbid it. Further facts confirm this view. In animals of a rather higher grade of organization, such as the more complicated worms, crustacea, some insects, and some molluscs, the power to regenerate is still present, but it is limited. No longer can the animal be cut in half, or into a number of little bits, without losing its capacity for life; but so long as certain central and essential organs remain at work, very considerable losses, such as those of a limb or a tail, can be replaced.

This degree of regenerative power continues even into the land vertebrates. Although the frogs and toads do not, the lower or tailed amphibia still possess it. Even a mature newt or axolotl can grow not merely a new toe or new foot, but can repair the loss of a whole hind limb and limb-girdle. This we may call organ-regeneration. In such animals, the early stages in development of an organ such as a limb are capable of complete regeneration.

Thus, by cutting a tadpole's limb-buds so that a number of separated bits are left, the animal may be made to grow more than the normal complement of legs. The same limitation of regeneration is seen, as might be expected, during development also. For instance, while the amphibian in its segmentation and blastula stages has an almost complete regenerative capacity, the tadpole can only regenerate single organs, and the frog can do no more than heal wounds. Even in newts and salamanders regeneration is much more rapid in the tadpole than in the adult. Among reptilia, lizards are the only animals which possess even the power of organ-regeneration, and in them it is confined to the tail.

When we look closely we find that here the original general power of regeneration inherent in living things has been retained in this organ for particular biological reasons, and has been combined with special adaptations to make it of greater value to the animal. The lizard has the power of self-mutilation (or *autotomy* as it is often called). If you catch a common lizard by the tail you will find that after a moment or two of squirming, the tail, broken off short at the root, will be all that you have in your hand, and the rest of the lizard will be running off into safe hiding. This is only made possible by a special structure of the vertebrae of the base of the tail; they have a crack on either side which penetrates almost to their centre, and the tailmuscles are so arranged that when one set of them contracts they pull at the further half of one of these vertebrae, and snap it off across the plane of weakness. The tail when broken off is, though doomed to death within a few hours, not yet dead; and it continues to execute the most violent wrigglings and jerkings for some time.

In Nature what occurs is apparently as follows.—When a bird or other enemy makes a pounce at the lizard, it will often only succeed in catching it by the tail. When this happens, the tail is broken off, and by its squirmings continues to keep the attention of the attacker while the lizard itself is making for safety. The tail is lost and devoured, but the lizard grows a new one. If lizard-eating animals were rational they would probably confine themselves to tails, gathering the annual crop as we gather the annual crops of fruit off a fruit-tree.

In any case, the lizard often saves itself by sacrificing its tail (*how* often could be established by finding out how long tail-regeneration takes, and then by examining a number of lizards from one locality and noting how many were in process of growing new tails—an experiment well worth trying); and this desirable end has only become possible by adding the special mechanism for breaking off the tail to the primitive power for regeneration. Crabs and many other Arthropods have this power of autotomy in their limbs, and some worms when handled break their whole body into fragments, each one of which will regenerate into a complete individual! One word about the power of regeneration in mammals and ourselves. Although we cannot grow new limbs, yet our wounds will heal, and this is still a real regenerative process; we may call it wound-healing or tissue-regeneration.

It is also to be remembered that in a certain sense the growth of such tissues as the epidermis is a continual regeneration—for something is continually being lost, and as continually being renewed. These last processes fall under the head of *normal* or *physiological regeneration*. A more striking example, perhaps, than the skin is to be seen in the antlers of deer—two masses of bone weighing several pounds which are shed and renewed each year. Two or three general principles emerge from these facts. The first—to recapitulate—is that regeneration is a primitive property of life.

Most low organisms never stop growing, and the capacity for regeneration is clearly associated with the capacity for growth, since growth is necessary for many of the processes of regeneration. Other changes that are at work in regeneration, however, have the character of rearrangements; for instance, in the formation of a whole newt out of one of the first two blastomeres of the newt's egg, the only process is one of regulation—of rearrangement of material which ought to have produced a half, so that it actually produces a whole of half size.

In the regeneration of a small piece of Planarian to form a whole worm, a certain amount of new growth takes place at the cut ends of the piece, but many of the changes involved are brought about by changes in the original piece. For this to happen, however, a certain degree of plasticity is needful—cells and tissues must be capable of altering their original character and adapting themselves to new conditions. This also is largely a property of actively growing organs; further, it appears to be impossible if tissue-differentiation has reached too high a pitch.

Thus the gradual limiting of regenerative capacity which we find as a general rule both in the development of the individual and in the progress of life as a whole, is due partly to increased differentiation, partly to the fact that the higher animals, instead of growing indefinitely, show a sharply marked stoppage of growth on reaching the adult state. Regeneration is thus essentially the restoring of typical form, structure, and function; we might call it a special case of regulation.

The fact that a piece of Planarian without a mouth can yet regenerate head and tail is obviously interesting. The piece's capital can be readily transformed into cash for the undertaking of new enterprises—in other words, the living framework of its body can be broken down into foodmaterials which can then be used up in new growth elsewhere. In the higher animals, the living capital is locked up in non-negotiable forms to a much greater extent, although not so completely as is often imagined.

This power of drawing upon the living tissues in case of need leads on to another remarkable power of Planarian and other low types. If an intact animal of this sort is starved, it can continue to exist for periods of weeks and months by thus liquidating its capital for its day-by-day expenses. Since it draws on its own tissues for its daily energy-needs, it therefore must grow smaller and smaller.

Planarians can continue this economical process until they have diminished their size to below that at which they hatched from the egg. If they are given food again (at any stage in the proceedings) they will start growing once more, and recover their normal size and appearance. One very curious fact was early noted about Planarians thus reduced in size by starvation—namely, that their proportions gradually altered from those of the adult and became more like those of normal young worms.

This led to the idea that perhaps they not only *looked* young, but had really *become* young again—an idea which was easily put to experimental proof. Some species of Planarians reproduce mainly asexually by transverse division; after the worm has reached a certain length a new head is formed a little behind the centre of the body, and in front of this a split appears along which separation eventually takes place. A brood of worms from one such species was taken. When they had nearly reached full size, they were divided into two lots. One lot, fed regularly, continued to grow and divide in normal fashion.

The other lot was starved until the average size of its members was reduced to about half; then it was fed until it had recovered its original size; then starved again—and so on repeatedly. The experiment was continued for a period of time in which the well-fed animals went through nineteen generations (a period which, though only occupying a few months for Planarians, would in human beings mean about six centuries —from Chaucer's time till to-day). During the whole of this period, none of the other lot reproduced at all, but all remained within the limits of size set for them by the experimenter, and, what is more, showed no signs whatever of age or of diminished vigour. In fact there is every reason to suppose that the experiment could have been continued indefinitely—in other words, that the individual Planarian, by proper treatment, can be made immortal.

The elixir of life was sought by the alchemists of medieval Europe for hundreds of years. These experiments show that it has at last been discovered; but, unfortunately, it is only effective with Planarians and some other animals, all of a low grade of organization. Before we go further into the principles involved in this matter, we must mention that there is another way in which animals may surmount starvation and other unfavourable conditions.

If an ordinary hydroid polyp like Obelia be kept in the laboratory, the hydranths, or separate organized individuals of the colony, will (unless the water is well aerated artificially) show a curious series of changes. Their mouths will close and their tentacles become stumpier; eventually they will lose almost all signs of their previous differentiated structure and become converted into a rounded or egg-shaped bag with a few little knobs marking where the tentacles had once been. They have lost differentiation—in other words, have undergone dedifferentiation.

In Obelia the process is complicated by the fact that while differentiation is going on, the cells of which the hydranth is made up detach themselves from their neighbours and, leaving their fixed place in the tissues, migrate into the central cavity and down into the common stalk which connects all the hydranths. By this means the hydranth dwindles rapidly, and is reduced soon after dedifferentiation is complete, to a tiny knob in the bottom of its protective horny cup. The animal has been resorbed by the stem. It is as if a house were to be unbuilt, by the bricks leaving their places in the walls and migrating into the garden. If, however, the hydranth is cut off at its base beforehand, the central cavity is soon filled up with detached cells, and after this, dedifferentiation alone occurs until the animal is reduced to a mere living bag tight packed with cells.

These types of dedifferentiation are found in many low types of animals, such as various Coelenterates and Ascidians. Among higher forms partial dedifferentiation may be a normal mode of development. For example, when the tadpole metamorphoses into the frog, some of its tissues start to dedifferentiate, for instance, those of the gills and tail. The process is complicated by the fact that phagocytosis, or the devouring of undesirable materials by white blood-cells, here reaches a far greater pitch of perfection than in lower animals, and, as a result, as soon as the cells have reached a certain degree of dedifferentiation, they are attacked and removed by the white blood-cells.

The disappearance of the tadpole's tail is thus due more to the activity of the phagocytes than to the migration of the tail-cells themselves. Similar methods of resorption of tissues are found when, for instance, a tumour or a graft disappears in a human being. To return to regeneration, and to another of its aspects, it is found that the presence of one part is often necessary for the regeneration of another part. In Planarians, for instance, the mouth opens at the end of a sort of trunk, the muscular pharynx, which can be protruded from the centre of the underside.

If a small piece of the flatworm is cut out by two transverse cuts, it depends on a number of circumstances whether a head shall be formed or not; for instance, a low temperature near freezing, or exposure to dilute alcohol or other poisons (the experiment is easy to carry out) will reduce the frequency of head-formation, or even suppress it altogether. If a head is produced, such a piece will always proceed to the formation of a pharynx.

But if no head appears, the formation of a pharynx depends on the original position of the piece in the body. If it was cut from in front of the original pharynx, it will often regenerate a pharynx, even in the absence of a head; but if it came from behind the original pharynx, a new pharynx will never be formed unless a head forms first. In other words, a new pharynx will be regenerated in the centre of a piece of Planarian provided that there exists at the front end of the piece a region which is normally anterior to a pharynx—whether this anterior region be an old part of the body or a new regenerated head. Thus, the head or front region of the body exerts some remarkable influence upon a distant region of the rest of the piece, making it construct an organ that it could not otherwise produce.

It makes things make themselves. Exactly how this effect is produced we do not know. Possibly it is through the agency of the nervous system. In any event, it is clearly a property of very great importance, and also one which seems to be of quite general occurrence. For instance, when the eye-stalk of a prawn is cut off, a new one like the old is usually regenerated. But when, in addition to the eye-stalk being amputated, the optic ganglion, or part of the brain to which run the nerves from the eye, is also cut out, the organ which is regenerated bears no resemblance to an eye and eye-stalk, but has precisely the form of part or all of the first antenna or feeler, the appendage which normally comes next behind the eye.

One other fact gained from the study of regeneration needs to be considered before the problems of normal development. If one of a pair of organs be removed, remarkable results often follow. For instance, in certain diseases, one kidney may have to be cut out. When this is done, the other begins to grow, and at the end of a few months weighs almost as much as the original pair together. This enlargement follows as a direct result of the increased demands made upon the functions of the surviving kidney, and is therefore known as compensatory functional overgrowth.

There is no regeneration of the missing organ, but a compensatory growth of its mate, resulting, one may say, in a *regeneration of function*. This functional overgrowth may be found not only to compensate for the loss of one member of a pair of organs, but also when part of a single organ is removed, or when an extra demand is made upon an organ without removal of anything. If, for instance, most of the liver or pancreas or reproductive organs are removed, the bits that are left will start to grow rapidly, until they are large enough to cope with the demands of the body. Or, as illustrative of the second case, the normal thyroid secretion is very rich in iodine.

When the amount of iodine in the food and drink falls below a certain quantity, the demands of the body for more iodine-containing secretion react on the thyroid; this begins to grow, producing more total amount of secretion as an attempt at compensation for its poverty in iodine. It may grow to a relatively huge size, producing a great swelling, called a goitre, in the neck. If a little iodine is given to a patient with such a goitre, the demand ceases, the reverse process sets in, and the swelling is reduced. A similar process takes place in muscles when extra demands are made upon them. Everybody knows how heavy exercise 'develops the muscles', and, after a preliminary period of loss of weight due to utilization of the reserves of fat, makes a man put on flesh.

The actual size of the muscles increases as the result of exercise. With this information we can now attempt some account of the normal development. It is found that the early development of a vertebrate such as a frog or newt falls, from our present standpoint, into three main periods. The first is essentially a period of the division and rearrangement of already existing materials, and takes the fertilized egg through segmentation and some way into germ-layer formation. The second is mainly one of differentiation—visible structure appears, the organs are blocked out, and tissues become different each from the other. Finally, in the third, the two most important features are growth, and the welding of the differentiated parts into a working whole—by mutual influence of organs upon each other and by an adaptation of separate organs to the demands upon them through the effects of function upon growth. This last phase may be taken first, since it is in many ways the most familiar. Think of a young organism—a recently hatched tadpole, a puppy, a child. Although it has already a characteristic organization, yet this organization is still plastic within wide limits. The final shape and size of its bones, its muscles, its sinews and tendons, its glands, its digestive tube, even of parts of its central nervous system, depend very largely upon the amount and the kind of use to which they are put during development. If one of a puppy's fore-limbs is tied up soon after birth so that it is never used, the dog will manage to get about very well with only three legs. But the leg which has not been used will be very different from the others.

The bones of its skeleton will grow almost to the normal length, but will never attain more than about half their normal thickness, nor will the arrangement of bony struts and stays at either end of the shaft (which in an ordinary bone are developed so as to meet both vertical and sideways strain in the mechanically most perfect way) be properly formed. The muscles of the leg, too, and their tendons will be very small. Not only this, but the nerve-cells of that region of the brain which controls the movements of the limb will be undersized. The power of bone to respond to the strains to which it is exposed, is of practical value. If, for instance, one of the small bones or phalanges of a finger is crushed and has to be removed, a small piece of bone from some other region can (if it include some of the periost layer from which new bone-cells are produced) be grafted in to take its place.

After a year or two the irregular splinter will have come to bear a marked, though not complete, resemblance to the bone for which it is doing duty. One of the most striking adaptations in the body is the fact that every tendon which attaches a muscle to a bone not only is of the proper size for the strain which the muscle can exert upon it, but is composed of parallel fibres which run precisely in the region of greatest stress. How can this beautiful relation of mechanism to function be supposed to have originated by natural causes? Does not the purposeful adaptation, different in detail for each separate tendon, demand the intervention of some supernatural power? Research, however, has in great measure removed such difficulties; for it has shown that all the separate detailed adaptations are the direct consequence of one inherent property of the tissue of which tendons are made. If a piece is cut out from a long tendon, such as the Achilles tendon which passes from the muscle of the calf over the heel to be fastened to the sole of the foot, it will after a time regenerate. Connective tissue grows out into the wound, the cells elongate and form fibres in the direction of the tendon, make attachment to the cut ends, and reproduce a structure essentially like that which was originally present. But if by one means or another the calf-muscle is put out of action (as when its nerve has been severed) then, although the gap will be filled by a new growth of connective tissue, the fibres will not become arranged in any particular direction, and no tendon-like structure will be formed.

Further, when in such an animal (the main experiments were carried out in rabbits) a piece of clean silk was healed across the middle of the wound, at right angles to the original tendon, and then, by winding it up very slowly on a screw, increasing tension was put on one end of it, it was found that fibres were formed in the new connective tissue, but parallel to the silk, and therefore useless for any restoration of normal function. From these facts (supported by many others) it is clear that tendon-fibres are formed from connective tissue when the connective tissue is frequently stretched, and further, that they are formed parallel to the stretching force. This stretching force is usually the force of muscular contraction (in growing animals there will also be found stretching due to growth of bones); so that tendons will be formed from ordinary connective tissue when muscles begin to contract, and will be 'adapted' to the direction and amount of stretching exerted by the muscle. In any case, all the apparent design seen in the adaptation of each several tendon to the demands made upon it, is apparently a direct result of this one reaction of connective tissue to tension.

To sum up, we may say that certainly most of the tissues of the body—possibly all—respond either throughout life or at least during growth (while their cells are still capable of multiplication) to the demands made upon them, and respond on the whole very advantageously from the point of view of the whole animal. The skeleton, the muscles, tendons, many and probably all glands, the central nervous system—all these we have already seen to show functional adaptation. So does the skin. Everybody knows that (after the blister stage has been negotiated) continual rubbing will make the skin thicken. 'Horny-handed' is so hackneyed an epithet for manual labourers as to have almost become mere journalese, and every oarsman demonstrates the fact in his own person. So, too, the circulatory system responds to functional demands.

If Everest is ever climbed, it will probably be because men can become 'acclimatized' to high altitudes and consequent low oxygen supply, largely, it would seem, through an extra production of red bloodcorpuscles by the bone-marrow and in consequence a greater capacity of the blood for capturing and transporting what oxygen there is. Furthermore, the size of a blood-vessel, both diameter of cavity and thickness of wall, is determined very largely-perhaps wholly—by the amount of blood which is pumped through it, and the pressure at which it is pumped. Cases are known in which large arteries become blocked up. If the blocking-up process is not too rapid, one or more small branches of the vessel will gradually enlarge, and produce channels capable of transporting the requisite amount of blood without difficulty.

This has been known to happen even when it is the human aorta, the main and indeed the only large artery springing from the arterial side of the heart, which has been blocked. The same kind of adaptation is also true of the protective reactions of immunity. It is a familiar enough fact that children who have had measles once do not generally catch it again, but this is only one among innumerable instances, such as the over-production of diphtheria antitoxin by horses injected with small doses of the toxin or poison produced by diphtheria germs, which has reduced the disease from a cruel scourge to a serious unpleasantness, or the benefits given by vaccination.

All these again depend on one single property of higher animals; when unaccustomed proteins enter their system, they can (if the foreign substances are not in too great amount, and provided certain other conditions are fulfilled) not only destroy or inactivate them, but in the process generate an excess of the destroying substance or *antibody*, which remains for a longer or shorter time so that a second invasion can be more rapidly and effectively crushed.

Although many of these adaptations are possible throughout life, others can only occur during development, and most of them are essential for normal development, since only through them do the size and structure of the various organs come to correspond to their function and to the physical and chemical stresses imposed on them. There is, however, another type of mechanism which is active through the later periods of development and plays an important role not only in this functional adaptation but also in fusing the parts of the organism into a more unified whole, and that is the endocrine system or sum total of the ductless glands.

Many of them, such as the thyroid and the pituitary, have indispensable functions as regards proper growth. The thyroid regulates the rate of metabolism, increased amount of thyroid secretion producing an increase of oxidation in the body. As might be expected, this function seems to be of more importance to the well-being of animals such as mammals, with a normal high metabolism (great heatproduction and constant high temperature), than to those like fish or frogs, which do not have to be constantly generating large quantities of heat to keep their temperature above that of their surroundings. Within the mammals, the absence or failure of the thyroid makes more difference to an adult man than to an adult sheep: the thyroidless sheep is but slightly abnormal, while the thyroidless man becomes notably fat, sluggish, and slow-minded.

But its absence in the child, in whom growth and activity should be much greater, makes far more difference again than in the grown-up man; the thyroidless human child or *crétin* usually dies early, can only grow into a stunted dwarf, and never develops normal intelligence, the brain of man apparently requiring a high level of metabolism for its unfolding. In Amphibia, metamorphosis from larva (tadpole) to adult is under the control of the thyroid. Tadpoles with their thyroid cut out refuse to be transformed into frogs, and grow to giant size in the water, while if a dose of thyroid is given to quite young tadpoles, they will metamorphose into froglets no larger than flies. The changes which take place in body and mind at puberty are also under the control of the endocrine system. The immediate control is exerted by the gonad itself. It is important to find that a whole set of separate changes can be brought about by a single change in the ductless glands, and further, that all these changes are in the long run related to one function, that of reproduction.

That a single original alteration in the chemical composition of the blood, an alteration appearing inevitably at the end of a definite span of growth, should bring about an increased growth-rate, the formation of mature spermatozoa in the testes, the appearance of hair on the face and body, and a marked change in mental outlook and temperament, is one of the most striking facts in biology. All the ductless glands have this in common, that their secretions are carried all over the body by the blood, so that they can exercise a number of different effects simultaneously upon the most diverse organs. It is fairly obvious, however, that for all the importance of function and of the endocrine system in development, their task is essentially one of regulating the degree of activity of processes already in being, or of directing them along particular channels. Bones exist in the absence of a pituitary or of pressure upon them, and muscles without being used.

It is only a small part of the adult man's growth of hair that is dependent upon the testis; metabolism continues in the absence of a thyroid, and the first development of a gland or blood-vessel takes place before its use begins. Another point that should not be forgotten is that in the long run the differences in shape and size between different animals depend almost entirely upon differences in the amount and direction of growth in different regions during development. In very early life the hind-limbs of a whale grow much less fast, compared with the body, than do those of a man, and those of a man much less fast than those of a kangaroo. Some idea of the changes which would be produced in human beings if the proportions between general growth in length and growth in breadth were rather different from what they actually are, can be obtained by visiting the distorting mirrors at a fair or circus.

Or one organ may be affected differently from the rest; the effect of this on general proportions may be seen by looking at oneself in the broadening mirror, first with the arms held against the sides, and then outspread at right angles to the body. If in the embryo of the fish Argyropelecus the tendency to growth in length had been a little less, and the main direction of dorso-ventral growth were not quite at right angles to the main direction of growth in length, it would have developed into a passable imitation of another kind of fish. No doubt such slight alterations in growth are at the bottom of much evolutionary change.

How do these more fundamental structures and functions originate to provide the primal basis to be shaped by the secondary moulding forces we have described? Although much is still not clear, yet the main outlines of the answer emerge when we look at the next previous stage of development, the stage of differentiation. This starts during gastrulation, and ends as the various organs become capable of exercising their functions. Its great characteristic is the sudden appearance, in the hitherto comparatively simple germ, of a number of chemically different regions, each one of which is irrevocably destined to generate one particular kind of organ.

What is more, this will usually be accomplished whether the developing organ-rudiment is in its proper relation to the rest of the organism or not. If the late gastrula is cut in two so as to separate front from hind region, each part (after healing the wound) will proceed to develop just those organs which it would have done if the embryo had been left intact. We are presented with an anterior half-embryo without any of the hinder organs, and a posterior half-embryo with tail, hind limb-buds, and most of the trunk, but no head-regions, gills, or heart. There is, in fact, in this stage *no* regeneration whatever of missing parts; yet, in spite of this, both halves appear quite healthy, and only begin to degenerate at a stage when the organs should be beginning to work and the animal to fend for itself. Or a piece of one embryo may be grafted on to another embryo, even of a different species, and both will continue to develop. One actual 'chimaera' has been artificially produced consisting of the front half of one kind of frog and the posterior half of another, which grew and metamorphosed quite happily.

Or if a small part of the future brain-region be cut out just after the gastrula stage, and grafted in again in reversed position, it will produce what it would have produced in an untouched embryo, although now of course the organs to which it gives rise will be misplaced. The eye, for instance, which is formed by an outgrowth from the brain, may by this means be made to develop behind the ear, and yet have normal structure. This will happen in spite of the fact that no visible trace of even the earliest eye-rudiment was present when the operation was performed. Before gastrulation, on the other hand, no such irreversible changes, inevitably determining the future appearance of organs, have yet set in.

Cutting the developing egg in half in the earliest stages has a totally different result from cutting it in half after gastrulation has set in. In the former case, as we have seen, one or both halves may become reorganized into a whole, and produce an embryo normal in every respect save size. One very remarkable fact has recently been discovered, namely, that the sudden change that takes place in a few hours during gastrulation in an animal like a frog or newt, during which the germ loses its capacity for regulation and becomes a sort of jig-saw of separate parts, each predetermined to produce just one particular organ of the future tadpole, occurs under the influence of one small region of the embryo.

This is the region just above the dorsal lip of the blastopore. By ingenious microscopic operations, even tiny bits of a newt's egg can be removed and grafted into other places on the same or other embryos. If this dorsal-lip region be taken and grafted into the flank region of another developing egg, it is found to cause the development of an extra set of organs, in addition to the one produced in relation to the host egg's own dorsal lip. An extra nerve-tube, notochord, muscle segments, kidney tubes, and gut may be thus induced to form—in other words, the beginnings of a whole second embryo. What is more, they are not formed by the grafted piece itself, but in the host-tissues near it. Again, we are reminded of Mother Carey in Kingsley *Water Babies,* who 'made things make themselves'. No other region has this power, and so the title of 'organizer' has been bestowed upon the dorsal-lip zone.

It is interesting to recall that a similar sort of power was exerted by the head of a Planarian in inducing the formation of a mouth and pharynx. Another fact is that the dorsal-lip region at this period is the most actively working part of the embryo, with very rapid cell-division. It seems probable that this great activity has something to do with its organizing capacity. Other interesting results, also apparently concerned with differences of activity in different parts of the developing egg, have been obtained with still earlier stages.

Although the period up till early gastrulation is of course an essential link in the developmental chain, no important steps in the construction of the details of the animal's typical organization are being taken so long as it lasts. The original comparatively simple organization of the zygote is not much altered; and the most essential change is the cutting up of the huge and unwieldy single cell, the fertilized ovum, into hundreds of small cells. And then each is a relatively independent unit, with its own cell-membrane as frontier, so that it is much easier to localize a number of separate chemical processes in different parts of an embryo divided up into cells than it would be to localize them if the embryo were all one large cell, in the same sort of way that the possession of a number of small pots and pans makes easier the task of preparing an elaborate dinner.

To choose another parallel from human affairs for the second and third phases of development, we might say that the second corresponds to the discovery and invention of new processes, new machines, new products, while the third corresponds to the growth of the industries based on these discoveries, a growth which is regulated by the laws of supply and demand, by the customs and laws of the country, and by the state of world affairs, but can only take place once the new inventions have been made. In conclusion, it is worth mentioning that every cell of the body appears to receive the same complement of chromosomes, the original set possessed by the fertilized egg being exactly duplicated at each cell division, so that each cell has a 'copy' of it. This at first sight seems to make differentiation hard to understand; but perhaps a musical analogy will make it clearer. The performance of a piece of pianoforte music demands a pianoforte. But we might have a hundred identical pianos in a row, from each of which would be emanating a wholly different piece of music: that would depend upon the players. For the pieces of music substitute the different organs of the embryo; then the pianos are the chromosome-sets and the players are represented by the different conditions in different regions.

For instance, at the close of segmentation some cells are small and poor in yolk, those at the opposite end are large and yolky, some are at the surface, others tucked away in the interior; and the differences between different parts are increased after gastrulation. In animals progressive development generally comes to a close with the attainment of the adult condition. In man, however, this need not be so in respect of one character. Mental development may continue; in other words, the contents of the mind may increase and its organization become more complex and more perfect, so long as maturity continues and senility has not set in.

Of course this is by no means always very marked, and equally of course something of the same kind is seen in the higher vertebrates which can profit by experience. But the extent of the possible development open to man is none the less very striking. One of the chief biological differences between man and all other organisms is that, through his power of thinking and learning, his development and differentiation can continue through the whole of life.

EVOLUTION OF PHYSICAL STRUCTURE

Birds and mammals both possess a constant temperature

which is ordinarily well above that of their surroundings. They should, therefore, be equipped with an arrangement for preventing too rapid loss of heat; and this is provided by the feathers in one case, the hairs in the other. Swimming birds, almost without exception, are webfooted or lobe-footed.

The Jacana, which seeks its food on the floating leaves of water-plants, has enormous toes which distribute its weight as do skis or snow-shoes on the thin crust of snow. Indeed, in general the feet of birds are remarkably well adapted to the life which their owners lead. The egg-eating snake, Dasypeltis, possesses a special mechanism by which it can temporarily dislocate its jaws to swallow an egg whole.

In addition, one of its neck vertebrae bears a downwardly directed spine which protrudes into the gullet, and is used to crack the egg where none of its contents will be wasted. The larvae of crabs live near the surface of the sea, and are provided with long spines which increase friction and make them less ready to sink. But the most striking examples are those known as convergence, where a similar mode of life produces similar effects on quite unrelated animals.

During the interval, progress has been steady and gradual in both departments of naval warfare, each advance in efficiency of guns being the stimulus for new invention in the methods of protection, and vice versa. It is when there is a general increase of the animal's powers of control that we speak of progress; when the increase is in one special particular only, we speak of specialization. For instance, the horse is specialized for running, the mole for burrowing, the bat for flying, the whale for a marine life, the lion for catching and devouring large animals, the sloth for living in trees.

Each of these animals is well adapted for its particular mode of life, but each is by that very adaptation quite cut off from leading the life of any other. During the late Secondary period there existed similarly specialized types of animals. For example, the Ostrich Dinosaur was adapted for running, the Ichthyosaur for life in the sea, the Tyrannosaur for preying on large animals, the Pterodactyl for flight, and so on. But all the former list of animals are mammals, all the latter were reptiles.

The mammals are higher in that they possess proper temperature-regulation, for instance, and better prenatal care of their young, as well as in many other points. Thus, while the types of specialization, or of adaptation to particular modes of life, are somewhat similar in the two cases, yet each member of the first group is higher than any member of the second, because its general organization is on a more efficient level.

Whenever a new group of animals is evolved, it is found that its members soon become specialized in different directions, thus filling up different vacant places in the economy of Nature. This adaptation to different modes of life, while, as we have seen, we call it *specialization* when we are thinking only of one species of animal, is called *adaptive radiation* when we are thinking of the group as a whole. All fish, for instance, breathe by gills, possess fins as limbs, and have other characteristics in common, so that any member of the group can be at once recognized by a brief examination of its structure.

Yet the detailed form of different species of fish is extremely varied. Besides the ordinary type of active, free-swimming fish, like the herring or trout or mackerel, there are fish flattened in adaptation to life on the bottom, some flattened sideways, like the sole and plaice, others flattened from above downwards, like the skates and rays; there are elongated fish like eels and pipe-fish; there are fish with prehensile tails, like the sea-horse; there are the flying-fish adapted for leaping long distances out of the water; mottled fish of irregular outline adapted for living on rocks; deep-sea fish with wonderful phosphorescent organs for searchlights and huge eyes for perceiving the faintest trace of light; cave-fish without eyes at all; Sucker-fish adapted for sticking tight to the underside of stones, or for being carried about by larger fish without expending any energy themselves; and so on and so forth through almost every conceivable sort of form possible in an under-water existence.

The examples previously mentioned give an idea of some of the adaptive radiation which has taken place in reptiles and in mammals; very similar instances could be taken from any other large group, such, for instance, as the insects. It is interesting to take any such group and to see what part adaptive radiation has played in giving rise to the main sub-groups into which it is divided. There is one particular form of specialization which we have not so far mentioned; and that is the retrograde form of specialization known as degeneration.

There are many cases known where animals can be definitely shown to be less highly organized to-day than were their ancestors in the past. The animal known as Sacculina, for instance, is a parasite upon various sorts of crabs. It consists of a mere bag filled with little else but reproductive cells, and sending out a whole series of branched roots which penetrate the crab's body in all directions and suck nourishment out of it. At first sight, the relationships of this unpleasant creature are very difficult to determine.

But when its development is investigated, it is found that it hatches out of the egg as a a free-swimming larva exactly like those found in many Crustacea. It has jointed limbs, an external skeleton made of chitin, and in fact bears all the distinguishing marks that other Crustacea do, be they crabs or lobsters or shrimps or water-fleas. It is, in fact, a crustacean which has become adapted to a parasitic life; and in so doing it has acquired special adaptations, such as the root-like organs, specially suited to that life, while it has lost sense-organs, limbs, digestive system, and everything else necessary for leading a free and independent existence.

It has lost more than it has gained, its organization has become simpler, its independence less; in fact it has gone down the evolutionary hill, and the direction of its history has been in most ways the opposite of the direction which characterizes biological progress. The form in which Sacculina hatches out resembles in general many crustacean larvae; but it is particularly like the early stages of the animals known as barnacles. Every one who has been to the seaside knows what an acorn barnacle looks like—a little creature attached to rocks or piles, enclosed in a white shell, and capable of sending out of a slit in the top of the shell a regular sweep-net composed of a number of 'arms'—really appendages—with which it drags minute floating particles of food in towards its mouth. Even in the adult state a barnacle shows some resemblance to ordinary Crustacea, especially in its jointed limbs and chitinous external skeleton; the early free-swimming larva clinches the matter and gives complete proof, as in the case of Sacculina, of their crustacean nature and affinities.

The very close resemblance of the larva of barnacles to that of Sacculina points to an especially close connexion between the two sorts of animals; and as a matter of fact, the two are undoubtedly descended from a common stock. The barnacle is degenerate as well as the Sacculina, but it is not so degenerate. It still, for instance, possesses organs for capturing and digesting food. On the other hand, it has lost its organs of special sense and of locomotion. Further, it is not adapted to the same general mode of life as is Sacculina; it is not parasitic, but sedentary or sessile.

This settling down and becoming fixed is the other great cause, besides parasitism, of degeneration in animals; as would be expected, the degeneration due to a sedentary life is rarely so great as that due to parasitism, since the sedentary animal does not obtain its food ready-digested as do most parasites. It must not be supposed that because the general rule among animals is that time brings change, that therefore time *invariably* brings change. The common lamp-shell Lingula, for instance, has persisted without any appreciable change whatever in the structure of its shell for the prodigious period of time, certainly over five hundred million years, which has elapsed since the Cambrian epoch in the Primary period. Even when individual species have changed, the general characters of groups have often persisted with very little modification, as, for instance, those of dragonflies since the coal-measures, of shark-like fish since the Silurian. In some cases this may mean that for some unknown reason the species or the group has lost the power of varying to any considerable extent.

More often, probably, it so admirably fills one particular niche in the order of things, and a niche which stays the same throughout the geological periods, that it pays for the animal or the group of animals to stay as they are, leaving it to other groups or to other branches of the same group to colonize new niches and to progress towards fuller existence. To sum up, we may say that two main types of evolutionary change result in animals (and also, as a matter of fact, in plants) from the struggle for existence and the constant appearance of inheritable variations. In the first place, once a new type or plan of structure has been evolved, it undergoes adaptive radiation; in other words, there are developed a number of separate species all built on the same general ground-plan, but adapted to different and usually incompatible modes of life.

In the second place, new types and new plans are continually appearing as time goes on, and progress is marked by the fact that among the later-evolved types there is to be found greater complexity of organization, greater control and independence of environment, than among the earlier. It might perhaps be thought that specialization was often the same thing as progress. Specialization, however, implies close adaptation to one particular mode of life, while progress means greater general efficiency.

If we look into the actual history of animals in the past, we find that specialization in one direction involves the sacrifice of possible advance in other directions, and is a barrier in the long run to any but a quite limited degree of progress. As a result of its long course of specialization, extending for tens of millions of years through the better part of the Tertiary epoch, a specialization all tending towards greater efficiency in running and browsing, the horse stock has, it seems, cut itself off from the possibility of adapting itself to other modes of life—to a life in the water or in the trees, or to a carnivorous habit. There is a limit to the perfection which any line of specialization can attain. While the horse was growing larger, developing hoofs, reducing the number of its digits, another and wholly different type was being evolved in the person of man.

If it were not that horses are useful to man, and are accordingly domesticated, they would now be wholly or almost wholly extinct. The 'natural' enemies of the horse are the large carnivores. These are built on the same general plan as the horse—the mammalian—and indeed are the results of the specialization of the same plan in another direction. The same limits are thus set to them as to the horse stock. Before the advent of man, a state of equilibrium existed between the horse and its enemies, the latter not able to destroy the former entirely, the first not able to escape the payment of some toll to the second. But the horse came up against the wholly new biological conditions introduced by the new type, man; it was in direct conflict with man's cunning and tools and his habit of hunting in bands; still more important, it had to compete with the indirect effects of the new development, such as the settlement and cultivation and enclosure of the land. Against all this the horse was powerless. It could not develop far enough or fast enough to adapt itself to such sudden changes, and as a result it is becoming extinct as a wild species.

Over and over again the same thing happens, and the specialized representative of the old type, plant or animal, is extinguished in competition with the specialized representative of the new. For tree-like representatives of the horsetail type, which existed in the Carboniferous period, we have seed-bearing trees to-day; for pterodactyls, birds and bats; for dinosaurs, the large mammals; for the large early amphibians, the Stegocephalia, we have crocodiles; for the abundance, both in numbers and in species, of the larger mammals in the Pliocene, we have the multifarious activities of the swarms of man. The new type seems always to have arisen from some comparatively unspecialized branch of the old, and to have attained its pre-eminence by means of adaptations towards general instead of towards special efficiency.

Man has ousted the other mammals from their dominant position owing to the development of his mind. Through his particular type of mind he is able to deal rationally with the problems that confront him; tools, machines, tradition, civilization, and unexampled control and independence have been the result. The human mind is not merely adapted for solving one or two particular problems; it represents a *method* more efficient than any previously adopted for dealing with any and every problem that may confront an organism. Man's body is not highly specialized, and he seems to have arisen, through a monkey-like ancestor, from some unspecialized early mammalian stock like the Insectivores.

Nor did the early mammals show any signs of specialization. All the fossil mammals that we know of during the time when the great period of reptilian dominance and specialization lasted were small, primitive creatures, at first sight not likely to wrest the palm from their powerful rivals. Very similar chains of events may be seen taking place in the evolution of human machines and inventions. Take, for example, the history of transport. The most primitive method was the carrying of single loads by human beings or pack-animals. After that came the invention of vehicles in general and of wheeled vehicles in particular. The wheeled vehicle became specialized ('adaptive radiation') in innumerable ways.

We have the war-chariot; the rapid vehicle for passenger transport and for pleasure; the heavy wagon and cart; the van and pantechnicon; the four-inhand mail-coaches bowling daily at fixed hours along the main roads of the land. A limit was set to the capacity and the speed of such vehicles by the speed and strength of beasts of burden on the one hand, and by the imperfections of road-surface on the other. At the beginning of last century a new type of vehicle was evolved for which these limitations no longer existed. It was discovered how to replace the energy of animals by that of steam, and in large part to overcome the difficulties of surface friction by making the wheels of the vehicles run on metal rails.

As a result, steam locomotives became for certain purposes the 'dominant type' of vehicle within an extremely short period of time. Here again the type itself has been improved, so that we have now for some time been close to the limit of its possibilities. It does not seem possible to run a profitable service at speeds of much over sixty miles an hour. About half a century later a new plan was evolved. The internal combustion engine was produced, and gave certain great advantages, notably in being ready to start at once without the long preparation of 'getting steam up'. It appears to be the fate of new types to lead a precarious existence for a considerable time before they can compete successfully with the dominant types of the period. This was so, as we saw, with the earliest mammals; it was so for the steam-engine; and it was so, to a very marked degree, for the internal combustion engines. They were laughed at when their inventors took them out on the roads; the law laid down that they should be preceded by a man with a red flag; the early defects in construction did actually occasion many a breakdown.

But within thirty years they came into their own. Meanwhile, still another competitor is in the field—the flying machine, a totally new type, abandoning not only the particular device of the wheel, but the whole element to which the wheel was adapted. It looks as if in certain respects, where speed is the main object, the aeroplane would become the dominant type of vehicle; but that it would leave the major part of transport to be dealt with by train and by motor. Another interesting parallel with the evolution of organisms is found in the fact that although there has been progress, although the dominant type of vehicle has altered with the passage of time, yet many representatives of the old types have survived. They have managed to survive by becoming restricted to a few special conditions and places. Packanimals, for instance, while once universal, are now only employed in mountainous or roadless countries.

The American buggy is still of the greatest use over the unmade roads which are still to be found in so many parts of the United States. The horse-drawn vehicle will long hold its own in businesses in which not much capital is available, and speed and great power are not of the first importance. This survival of all or almost all the types that arise in evolution, even though new types arise and supremacy changes hands, is of general occurrence in the development of animals and plants, and is at first sight very puzzling. However, the insight which is gained by looking, as we have done, into the evolution of something familiar and human like the means of locomotion, helps us to understand the more complex and slower-moving processes of organic evolution. Another point which is brought out by the study of the development of some human contrivance such as the means of transport, is the great speed of change now possible in human affairs as against the slowness of change prevailing in lower organisms. The whole period from the stage-coach to the aeroplane is comprised in well under two centuries. The resulting change in human habits has been enormous; to produce comparable changes in the habits of an animal stock there would be needed a period certainly to be reckoned in tens of thousands, possibly in millions of years.

On the other hand, the evolution of machines is a perfectly real evolution. Two different types of machines capable of performing the same general function—such, for instance, as the motor-lorry and the goods steam-engine—do come into a very real competition with each other, and the issue of the struggle is decided by a form of true natural selection, depending in the long run upon which of the two pays the better. Here again the study of machines throws light upon the course of events in animals. It is often supposed that evolution must involve some conscious effort on the part of the evolving organism, that the struggle for existence is a conscious struggle, or that a species in some mysterious way 'learns' how to develop some new improvement in its structure. As a matter of fact, almost the whole of such ideas are purely metaphorical, and arise simply because we read the processes of our own minds into the operations of nature; it is not scientifically correct to speak, for instance, of *purpose* except in relation to human minds.

We see at once that the machines have no idea themselves of the direction of their evolution, that the 'struggle' between them is only a metaphorical struggle, that the selection between them is so far as they are concerned a mere sifting process, the issue of which depends upon the advantages or disadvantages which they may happen to possess. It is the same with organisms. If two races of animals come into competition, the issue is decided by the qualities which each happens to possess; 'natural selection' is a name for the effect exerted by all the forces of the environment with which they come into relation, an effect which acts again like an automatic sieve, and lets some through to perpetuate themselves, keeps back and so extinguishes others. The 'struggle' and the 'competition' are again usually metaphorical only. For instance, when the common housesparrow was introduced into America, it entered into competition with many of the small sparrow-like birds which had developed in that country. But the struggle did not take the form of a war between the invader and the original inhabitants.

It was an indirect struggle, due to the fact that both lived upon the same sort of food, both occupied the same sort of sites. The European sparrow happened to be endowed with qualities which gave it an advantage, and as a result it has spread enormously over the American continent, while many of the native birds have correspondingly decreased. If we wish to use a human metaphor, we can say that its success has been the result of 'peaceful penetration', not of fighting. The difference between machines and organisms, of course, is that the machines are directly designed by man, whereas the place of the designer in animal evolution is (roughly) taken by the variation which seems to be universal in organisms.

It is variation which provides the raw differences upon which the sieve of natural selection can work. Mention must also be made of the theories of evolution which are summed up under the term *orthogenesis*, which means evolution in straight lines. It is frequently found, as for instance in the development of the horses or of the elephants, that evolution as revealed by fossils proceeds straight onward through geological time in a perfectly definite direction-in the horses towards single toes and hoofs, in the elephants towards great bulk, tusks, and trunk. Orthogenesis is sometimes used merely as a descriptive term, to denote this observed fact of straight-line evolution. But by others it is used to mean that there exists some inner necessity for the evolutionary line in question to develop in just that one way and no other. However, a series of fossils, even if beautifully complete, can really give us no insight into the method by which its evolution occurred. Whenever the direction in which the series is evolving is adaptive or seems biologically advantageous, the orthogenetic series can be perfectly well explained by natural selection. On the other hand, there do exist cases where at least no advantage can be perceived by us in the direction pursued. This is so, for instance, as regards the Ammonites, the extinct cephalopod molluscs which died out near the end of the Secondary period. Near the close of their time on earth, they often evolved orthogenetically into the most bizarre forms, their spiral shell becoming unwound, or irregular. Possibly in such cases a real causal orthogenesis, an inwardly determined mode of variation, is at work. But we are justified in saying that such cases, if they occur at all, are certainly rare.

3

Absorption and Assimilation in Animals

The organisms in which food is taken into an alimentary canal and then digested, the products of such digestion must pass through the walls of the alimentary canal in order to find their way to the tissues and cells. This introduces the question of intestinal absorption. The problem of the mechanics of intestinal absorption has had a fascination for many physiologists. In the study of such absorption there is a possibility of applying physicochemical knowledge.

Thus, it is obvious that digestive changes, which convert indiffusible materials such as proteins and starches into diffusible compounds of much smaller molecular dimensions, make possible the passage of these now diffusible compounds through the membranous lining of the intestine. But although the laws of diffusion, and the laws of osmosis, are important factors in explaining intestinal absorption, there remain facts which cannot be explained on any simple physicochemical basis.

To arrive at a complete understanding, either one must assume strange and relatively unknown physical processes, or one can, if one please, more or less beg the question by assuming vital or undetermined biological factors. In higher animals the lumen of the intestine is lined with finger-like projections or villi. These have muscle fibres in their walls, and they appear to be in a constant state of motion, now shortening, now lengthening. Each villus seems to contract independently, and in dog intestines the rhythm of contraction is at the rate of about 6 per minute. These contractions were first clearly noted by Hambleton in 1914. Since then they have been carefully studied by Verzár, and by various other workers. Contractions of the villi seem to have some significance for intestinal absorption.

If the movement of villi is stimulated, intestinal absorption increases. Thus if HCI is added, there is an increase in rate of movement and an increase in the rate of absorption of glucose. (The effect is believed to be due to a hormone produced in the intestinal wall by the acid.) Condiments such as pepper, paprika, onion and garlic also cause an increase in the rate of movement of villi. Conversely, if cocaine is added, the movement of the villi may be prevented and then absorption of glucose drops 12-30 per cent. It thus seems clear that the villi play an important part in the absorptive activity of the mammalian gut.

As an argument against the importance of the villi it has been claimed that in the case of animals without intestinal villi the same type of absorption occurs. This claim, based on widely quoted experiments of Cohnheim on the isolated intestine of sea cucumbers, is apparently invalid. In the study of intestinal absorption, food materials or chemical substances of one sort or another are inserted into a loop of the intestine, and then after an interval of time the contents of the loop are examined in order to determine how much water and dissolved substance have escaped from the interior of the loop. Either the animal is anesthetized and operated on and the desired portion of the intestine tied off, or an isolated portion of the intestine is permanently sewed to the abdominal wall (*Vella fistula*).

Both methods have advantages and disadvantages, but in both of them there is some tendency for the intestinal loop to be under abnormal conditions which may easily affect movement of the villi, etc. A third method is that of Cori. In this method a small amount of the substance to be investigated is fed by stomach tube to a small animal, *e.g.*, a rat. After a given time the animal is killed and the amount of substance remaining in the intestine is determined quantitatively. The difference between the amount fed and the amount recovered from the whole intestinal tract is the amount of substance absorbed. Although with Cori's technic it is possible to study absorption under normal conditions, there may be variation in the length of time the introduced substances remain in the stomach before passing to the intestine, and the time factor for absorption is therefore uncertain.

In the human, it is possible to insert tubes not only into the stomach but into the intestine as well. With the aid of such tubes samples of stomach or intestinal contents can be analysed at intervals following the introduction of sugars or other food substances. In this way intestinal (and stomach) absorption can be studied in a relatively normal subject, but it must be remembered that the movement of villi which is under nervous control may be influenced to some extent by any irritation produced by the insertion of the tubes.

There have been many studies of nutrition based on the use of isotopes, and these nutritional studies often give information concerning intestinal absorption. Deuterium (heavy hydrogen) has been widely used and also radioactive isotopes of phosphorus, sodium, etc. In these studies, which will be referred to again later in the chapter, the isotopes or compounds containing them are introduced into the intestine and their presence is then recognized in blood or tissues. Sometimes the procedure may be rather simple. Thus a man can be fed radioactive sodium and following suitable intervals of time the presence of radioactive sodium in his hands can be determined with the aid of a Geiger counter.

Quantitative study of intestinal absorption has been confined almost entirely to the vertebrates, and most of the work has been done on mammals. This is unfortunate, for comparative data on absorption rates in different animals might be of considerable interest. According to Jordan and Begemann the intestine of the snail (Helix pomatia) is more permeable to disaccharides than is the intestine of the frog. In some invertebrates (sponges, coelenterates, flatworms, echinoderms) intake of food from the body cavity or intestine is largely controlled by phagocytosis. This is not true for insects. As an example of an absorption experiment, the following table is taken from Höber. It represents an experiment of Heidenhain's on an intestinal loop of a dog. The ends of the loop were tied off, but the loop itself was provided with blood through the mesentery. The record shown in the accompanying table indicates that the membrane lining the intestine does not behave as a semipermeable membrane, and that the laws of osmosis are obviously not followed.

Thus, when solutions of sodium chloride more concentrated than the blood are introduced into the loop, they lose water and, moreover, sodium chloride passes out of the loop. Nevertheless, it is obvious that osmotic forces are not without effect, for the stronger solutions of sodium chloride lose much less water than do the weaker solutions.

Introduced Into Loop		Time For	Removed From Loop			
cc.	Percentage NaCl	Total NaCl, Gm.	Absorption. Min.	cc.	Percentage NaCl.	Total NaCl, Gm.
120	0.3	0.36	15	18	0.60	0.108
120	0.5	0.6	15	35	0.66	0.23
117	1.0	1.17	15	75	0.90	0.67
120	1.46	1.75	15	109	1.20	1.31

The rate of absorption of various types of dissolved substances from the lumen of the intestine has been studied by many authors. Some of the papers are referred to in the reviews of Goldschmidt, Magee and Verzár, and the monograph of Verzár and McDougall. Salts such as magnesium sulfate and sodium sulfate pass through the intestinal wall much less readily than sodium chloride; they cause a flow of water into the intestine and thus produce a laxative action. Höber and Höber have studied the absorption of various organic substances in solution. For substances of no physiological importance, rate of absorption decreases with rising molecular volume. However, substances of distinct lipid solubility are absorbed at a higher rate than those not soluble in lipids. There is thus a similarity to cellular permeability.

Höber and Höber found that amino acids were absorbed at a higher rate than non-physiological substances of the same molecular volume. Numerous authors have compared the rate of absorption of various simple sugars.

In the mammalian intestine generally the rates are in the following series: galactose 0 glucose 0 fructose 0 mannose 0 xylose 0 arabinose. A similar series has been described for the intestines of the pigeon and the frog, but in the cat intestines, xylose is absorbed as readily as glucose. This paper has many references to the literature. In general, the sugars most readily absorbed are those most needed in the body. According to Verzár and McDougall: "In the absorption of certain substances there is a special selective advantage which seems to be completely contrary to physical laws.

These 'favoured' substances are ones which are especially used in the metabolism of the body. Such a specific selectivity is the result of active life processes in the mucosal cells. Thus from a mixture of hexoses and pentoses, glucose and galactose are selectively and quickly absorbed (three to four times as quickly as the others from blood-isotonic solutions)." The phosphorylation theory proposes an explanation of the selective absorption of sugars.

It was suggested originally by Wilbrandt and Laszt that the reason hexose sugars were absorbed more readily than pentoses such as xylose (in spite of their greater molecular volume) was the fact that hexoses were able to combine with phosphate. In the cells of the intestinal wall such "phosphorylation" was believed to occur and to be responsible for the more rapid absorption of glucose. The phosphorylated sugars must be dephosphorylated again as the sugars pass into the blood from the intestine.

The evidence in favour of the theory has been many-sided. Wilbrandt and Laszt showed that the presence of phosphates in the intestine favours the absorption of sugars and that iodoacetic acid which prevents phosphorylation also prevents rapid absorption of the hexoses.

Similarly, according to Lundsgaard, the alkaloid phlorhizin prevents phosphorylation and also the selective absorption of hexoses. Removal of the adrenal gland is also believed to interfere with phosphorylation and with the absorption of sugars. The literature on phosphorylation is voluminous and not wholly concordant. The older papers are referred to in Verzár and McDougall's monograph, already cited. Objection has been raised that iodoacetic acid is a serious poison and does much more than merely prevent phosphorylation.

Moreover, there is some question as to whether phlorhizin is an agent which specifically inhibits phosphorylation in vivo. On the whole the evidence indicates that the phosphorylation reaction hastens the absorption of sugar. Kjerulf-Jensen and Lundsgaard attempted to study the rate of dephosphorylation in the intestine in order to discover whether phosphorylation-dephosphorylation reactions could be rapid enough to explain selective absorption. They injected glucose into the intestines of rats, killed the animals 15-20 minutes later with strong cyanide solution, and then inasmuch as the cyanide prevented phosphorylation, they were able to follow the course of dephosphorylation.

Phosphate esters present in the intestines of sugar-fed rats were able to release inorganic phosphate rapidly. Within one or two minutes the reaction went to completion. Kjerulf-Jensen and Lundsgaard conclude that the phosphorylation-dephosphorylation reactions in the rat intestine are rapid enough to account for the phenomenon of selective absorption of sugars, but they do not regard their results as of necessity proving that phosphorylation plays the essential role in the absorption of hexose sugars, for it is possible that the phosphorylation reaction may merely provide the source of energy.

In an earlier paper, Lundsgaard was not able to obtain much evidence for phosphorylation in the intestine of the cat. This may be correlated with the fact already noted that the cat intestine shows no selective absorption of hexose sugars. Ideas on fat absorption have changed with the years. Before 1900 it was commonly held that some of the fat taken into the body of an animal was absorbed only after it had been hydrolyzed into fatty acid and glycerol, but that in addition some fat could be absorbed as finely emulsified particles. In 1900, Pflüger argued strongly that *all* the fat had to be hydrolyzed in order to be absorbed and that no emulsified fat could be taken into the wall of the intestine. This then became the accepted belief and found its way into all the textbooks. In 1940 Frazer was impressed with the fact that fat could be absorbed in the form of finely emulsified droplets.

An interesting argument in favour of the absorption of emulsified droplets of oil is the fact that paraffin oil is not ordinarily absorbed, but if it is emulsified by the addition of small amounts of oleic acid and cholesterol, then it is taken into the intestinal wall. As Frazer found more and more evidence that emulsified fat could be absorbed, he came more and more to doubt whether any large amount of fat was absorbed in the form of fatty acid and glycerol. Thus he points out that higher fatty acids are insoluble in water.

Of course the salts of the fatty acids (soaps) are highly soluble, but Frazer states that at the slightly acid *p*H of the small intestine of higher animals, the amount of soap that can be present is negligible. (Actually, even at *p*H 6, there is an appreciable amount of fatty acid present in the form of soap.) The emulsification of fats is favoured by bile and by the lecithin that is typically present in bile.

Frazer believes that the tiny droplets of fat enter the cells lining the intestinal epithelium through pores present in the outer boundary of these cells; these pores lead to canals, and the canals lead to the naked protoplasm. As a matter of fact, it has been observed both for plant cells and for simple animal cells such as sea urchin eggs that if an oil droplet is placed in contact with the cell membrane the oil will rapidly enter the cell interior. Frazer may go too far in claiming that practically all fat is absorbed in emulsified form. No doubt some is absorbed in this way, but probably a very appreciable amount is also absorbed as fatty acid and glycerol. This is strongly indicated by an isotope study of phospholipid absorption made with the aid of radioactive phosphorus. In this study it was clearly demonstrated that some of the lipid was absorbed without hydrolysis having taken place, but that some of the lipid was absorbed after hydrolysis. A similar result was obtained in an isotope study of the absorption of triglyceride fat. Phosphorylation is also believed to be important for the absorption of fats.

The idea is originally due to Sinclair. He thought that fatty acids entering the intestinal wall combined with phosphate (and a base) to form phospholipids and were then split off again to form neutral fat. Much of the evidence for the importance of phosphorylation in fat absorption is similar to that advanced to show the importance of phosphorylation for carbohydrate absorption; it is reviewed in Verzár and McDougall's monograph, also by Bloor.

In a study with radioactive phosphorus, Zilversmit, Chaikoff and Entenman concluded that the formation of phospholipid was not a necessary step in the absorption of fat. In other words, fat can be absorbed even if no phosphorylation occurs. Concerning the mechanism of absorption and the forces responsible for passage of water and dissolved materials through the intestinal epithelium, there has been wide diversity of opinion. Goldschmidt cites 15 different types of theoretical explanation. After summarizing the evidence, Magee in his review states: "it will be seen that the majority of the experimental results cannot be explained by known physical laws.

The conclusion may therefore be drawn that the phenomena of absorption are due partly to some special property of the epithelial cell." One difficulty that all theories have to face is the fact that active osmotic work occurs during intestinal absorption. The energy for this work presumably comes from the protoplasmic activity of the intestinal cells. But what sort of activity is involved and how is the energy applied? An attempt to find some sort of answer to this problem was made by Visscher. He and his collaborators have carried out thorough and extensive investigations on the problem of salt absorption, and they have shown very clearly that other than osmotic forces must be involved. Ingraham, Peters and Visscher adopted what they called the "fluid circuit" theory. This theory maintained that an ion such as chloride passed in through one region of the intestinal membrane and that water passed out through another region. The driving force of such an arrangement is not clear. In their discussion of theory, Peters and Visscher stated that there were five general types of theory (the fluid circuit theory constituting the fifth).

They easily disposed of the first three types of explanation, and the fourth which postulates some type of compound formation, they attacked on the ground that no substance was known which would combine with NaCl. However, this is hardly a fair criticism, for the proteins of the cell certainly can combine with cations such as sodium, and there is also a possibility that they may combine with chloride. As a matter of fact Hanke and Donovan claimed that 35 per cent of the chloride of the intestinal wall was bound. In view of the fact that in living systems practically all the energy produced in the course of vital activity is derived from chemical reactions, it seems logical to assume that the energy required to transport dissolved substances into the intestinal wall against osmotic forces is essentially chemical.

Franck and Mayer have considered the problem from a thermodynamic standpoint and they have postulated an osmotic diffusion pump which would obtain energy from chemical reactions and would be able to transfer either solvent or solute across membranes.

In their discussion of the types of reactions which might conceivably be involved, Franck and Mayer consider both phosphorylation reactions and the combinations of ions with substances like resins. Of course there are no resins in ordinary protoplasm; but, as already pointed out, ions can and do combine with proteins, and with proteins as with resins exchange of ions is possible. The absorption of calcium presents a special problem and there is a large literature on it. This ion, so important in vital processes, is absorbed only with difficulty from the intestinal tract. Diseases such as rickets are due to faulty absorption of calcium. The fact that vitamin D favours calcium absorption can perhaps be explained in terms of the effect of the vitamin on the combining power of calcium with the proteins or lipids of the cell. This aspect of the subject needs further investigation. In the physiology of mammals and higher animals generally, there is not only the problem of absorption into the blood or lymph from the intestine, but also the problem of how the tissues absorb dissolved substances from blood or lymph. This latter question has received relatively little attention.

Cori and Goltz studied the absorption of sugar by the liver and muscles from the blood. The absorption rate was rapid and approximately the same for both types of tissue. The absorption of food materials by the cells of animals is to some extent paralleled by the absorption of food by the plant. In plants, generally, there is no necessity for digestion of solid food, and only in insectivorous plants does this occur.

But in so far as the intake of food into the cells themselves is concerned, both in plants and animals there is evidence of a selective action on the part of the living cells. In plants also, there may be an active transport of dissolved substances against a diffusion gradient. This is indeed a characteristic of many, perhaps all, types of living cells and is not confined to the cells which are concerned with the uptake of food in animals and plants.

ASSIMILATION

Let us now assume that the raw materials which the living cell requires for its activities have found their way into the protoplasm. In general, the food substances absorbed by cells are immediately built up into more complex compounds, and to some extent into protoplasm itself. This is not true for every substance taken into the cell, for water which passes into the protoplasm may retain its chemical identity.

Moreover, it is possible that simple sugars absorbed by animal cells may be oxidized to produce energy even before they are synthesized or bound chemically. However, it is broadly true that the chemical compounds taken in by cells are relatively simple compounds, and that these are then converted into more complex materials which may often be considered as essential parts of the protoplasm itself. This process is termed "assimilation." In considering the facts of assimilation in various types of organisms, there is a sharp contrast to be noted between the green plants on the one hand and metazoan animals on the other.

As is well known, the former are able to synthesize all the complex chemical materials of their protoplasm from simple inorganic compounds. In animal cells, at any rate in the cells of metazoa, such synthesis from inorganic compounds is generally not possible. Instead of attempting to consider all types of living material, we shall confine our attention to the cells of green plants and of metazoan animals.

ASSIMILATION IN GREEN PLANTS

Photosynthesis. Green plants are able to utilize the energy of the sun for the manufacture of carbohydrates. This process is called "photosynthesis." It is of tremendous importance not only for the nutrition of plants, but also for the nutrition of practically all other forms of life. In the last analysis, even those animals which prey on other animals are dependent on the plants which these captured animals have eaten, for it is only by photosynthesis (or by the activities of autotrophic bacteria) that the organic compounds on which animal life depends can be created.

The wealth of the world also depends largely on the existence of photosynthesis now or in the past for, in addition to farm and lumber products, coal, oil, and gasoline owe their origin to it. Practically all the energy which comes to the earth is obtained from the rays of the sun; some of this is the result of the evaporation produced by the heat of the sun, for such evaporation is responsible for waterfalls and hence waterpower, but a greater part is due to the ability of plants to utilize the light of the sun in photosynthesis. Thus, in spite of the fact that the capacity for photosynthesis is not a general attribute of all types of protoplasm, its importance for all organisms and the fact that it does occur in many of them are reasons sufficient for a discussion of it in a book on general physiology.

However, only a brief survey of the field will be possible. The literature on the subject is voluminous. The older papers are reviewed in the monographs of Stiles and Spoehr. There is also a more recent monograph by Baly which is concerned chiefly with Baly's own researches. A survey of many aspects of photosynthesis may be found in a book by Rabinowitch. Other useful sources of information are the symposium on Photosynthesis in Plants, edited by Franck and Loomis and published by the American Society of Plant Physiologists, and the symposium of the Society for Experimental Biology on Carbon Dioxide Fixation and Photosynthesis.

Many reviews have also been written. In addition, various texts on plant physiology may also be consulted. In the process of photosynthesis, carbon dioxide and water are combined to form carbohydrate.

In view of the fact that carbon dioxide and water contain more oxygen than the sugar or starch which they produce, free oxygen is a by-product of the reaction, and is given off as a result of it. The carbon dioxide is obtained from the air or, in the case of aquatic plants, from the gas dissolved in the waer. In the purple and green sulfur bacteria photosynthesis occurs, but although carbon dioxide is involved in the synthesis, there is no release of free oxygen.

Methods of Measurement

There is one difficulty common to all methods of measurement. Whereas in photosynthesis carbon dioxide is taken up by the green plant cell and oxygen is given off, during respiration of the cells the very opposite occurs and oxygen is taken up and carbon dioxide is given off. Cells of green plants show normal respiration in the dark, and it is generally believed that this respiration continues in the light.

Therefore, in determining the source of carbon dioxide for photosynthesis, one must include not only that provided by the environment, but also that produced by respiration. Similarly with regard to oxygen, not all that evolved in photosynthesis is set free, but some of it is consumed by respiration. In the older literature various approximate methods were used to measure the amount of oxygen liberated in the photosynthetic process. These methods included the counting of the bubbles of oxygen liberated from a plant submerged in water, the use of motile bacteria and luminescent bacteria as indicators for the presence of oxygen, and the detection of oxygen by its effect on the leukobases of various dyes.

These methods are now chiefly of historical interest. At the present time, the student of photosynthesis has his choice of a number of exact methods which have been developed for the rapid determination of small amounts of oxygen and carbon dioxide.

Such methods, most of them developed by students of respiration, include the use of manometer technics and various other procedures. Franck, Pringsheim and Lad have used a photoelectric method for determining small amounts of oxygen. They measured the quenching of phosphorescence of the dye tryptaflavine adsorbed on silica gel. The method is useful for the study of photosynthesis under anaerobic conditions. In addition to methods involving determination of oxygen or carbon dioxide, the course of photosynthesis may be followed by determining the increase in the end products. Two methods are generally used in such study. Sachs estimated the gain in carbohydrate by determining the change in dry weight of a green leaf. The results obtained in this way have been shown to be rather inaccurate. Another method is to determine the sugar content of leaves before and after photosynthetic activity.

FACTORS WHICH INFLUENCE THE RATE OF PHOTOSYNTHESIS

The rate of photosynthesis is influenced by a number of factors, some of which are inherent in the cells of the plant, others in the external environment. Stiles lists a total of 15 factors.

Of these, some are absolutely essential for the occurrence of any photosynthesis at all, others merely determine the rate of the reaction. The existence of so many factors was a source of confusion in the older literature. Later workers have in general realised the necessity of keeping constant all variables but the one which is being especially investigated. *Concentration of Carbon Dioxide.*

In the complete absence of carbon dioxide no photosynthesis occurs; thus the presence of some carbon dioxide is an essential factor. This can be demonstrated by the addition of calcium hydroxide to the water surrounding green plants. When this is done photosynthesis ceases. For low concentrations of carbon dioxide the rate of photosynthesis is proportional to the concentration of carbon dioxide. Such a relation appears to hold good for plants living in air as well as for those living in water. A careful study is that of Warburg on the green alga Chlorella. In this form the rate of photosynthesis is closely proportional to the carbon dioxide present within the limits of 0.05 to 10, where unit concentration is taken as that of carbon dioxide in water in equilibrium with that in the atmosphere.

Above this upper limit, further increase in carbon dioxide concentration results in smaller and smaller increases in photosynthesis, until the latter remains constant in spite of any further increase in carbon dioxide. Apparently at this point some factor other than carbon dioxide is limiting the rate of the reaction. This is in accordance with Blackman's principle of limiting factors, which states that when a process is conditioned as to its rapidity by a number of separate factors, the rate of the process is limited by the pace of the 'slowest' factor."

Blackman's principle has had a wide influence on the literature of photosynthesis. Nevertheless, it is probably not rigidly true, for the rate of photosynthesis seems to be affected not only by the slowest limiting factor, but by others which may be nearly as slow. At very high concentrations of carbon dioxide there is a decrease in the rate of photosynthesis.

Light

Like the presence of carbon dioxide, light is an essential factor, and by definition no photosynthesis can occur in the dark.

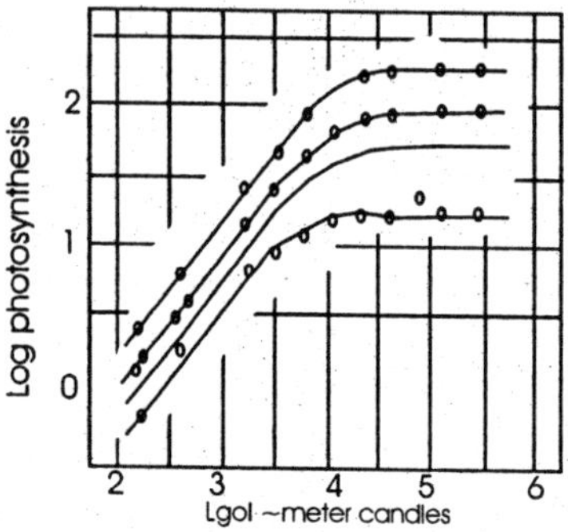

Fig. The Effect of Light Intensity on the Rate of Photosynthesis in the Fresh Water Plant, Camboda.

Photosynthesis is in terms of cu. O2 per hour per 100 mg. wet weight of Tissue. The Photosynthesis Scale is correct only for Curve *A*; the others have been Shifted Downward.

Both the intensity of the light and its wave length are important. As the intensity of the light is increased there is at first a proportionate increase in the rate of photosynthesis; then, with higher intensities, the curve tends to level off just as it does for higher concentrations of carbon dioxide. This general result has been obtained by various authors. More recently it has been demonstrated that very strong light causes a decrease in the rate of photosynthesis. In comparing light of different wave lengths it is essential to use radiation of the same intensity.

Older investigators often failed to do this. For higher plants, most authors find that the efficiency of the photosynthetic process becomes less with decreasing wave length. Gabrielsen himself in a study of four higher plants found that if the photosynthetic rate in the red region of the spectrum (650 mì) is taken at 1, that in the green 545 mì) is 0.62 and that in the blue (440 mì) is only 0.37.

Similar figures were found for the alga Chlorella by Warburg and Negelein. On the other hand, in studies of marine algae such as Ulva, some authors find two maxima of photosynthetic activity, one in the red, the other in the blue. Similarly Hoover finds for young wheat plants that there is a maximum at 650 mì in the red and a somewhat lower maximum at 440 mì in the blue and that the intermediate green region is less effective in promoting photosynthesis.

Chlorophyll

The presence of chlorophyll is also an essential factor in photosynthesis, and the process does not occur in plants which lack chlorophyll. Within the plant cell, the chlorophyll is contained in structures known as chloroplasts. These typically are discs or flat ellipsoids, measuring about 3-10 ì across. They may contain other pigments in addition to chlorophyll. There are two main types of chlorophyll, chlorophyll a and chlorophyll b. The former is believed to be $C_{55} H_{72} O5N_4 Mg$; the latter $C_{55} H_{70} O_6 N_4 Mg$.

Physically, and to some extent chemically, chlorophyll is a wax. On hydrolysis it yields two alcohols, methyl alcohol and phytol ($C_{20} H_{39} OH$). Various chemists have attempted to determine the structural formulae of the chlorophylls. Fischer has suggested two possible formulae, one of which is shown in figure (the other differs only slightly). It will be noted that Mg is at the centre of the molecule and that it is surrounded by four pyrrole rings.

Chlorophyll *b* differs from chlorophyll a in that at the no. 3 position (pyrrole ring II), the CH_3 (methyl) group is replaced by a CHO (formyl) group. Various authors point out the similarity in the chemical constitution of chlorophyll and hemin, which latter forms part of the hemoglobin molecule. That both hemoglobin and chlorophyll are involved in reactions concerning oxygen and carbon dioxide seems more than a coincidence. The hemin molecule also contains four pyrrole rings, but at their centre there is an Fe atom instead of a Mg atom. Like hemin, chlorophyll combines with protein. On the average chlorophyll *a* is three or four times as abundant as chlorophyll *b*.

Chlorophyll *a* and chlorophyll b are not the only pigments found in plant chloroplasts. In all, there appear to be over 60 such pigments. The subject of these pigments is discussed at some length by Strain. Chlorophyll *c* is found in brown algae, in diatoms and in dinoflagellates; chlorophyll *d* is found in red algae. The chemical composition of these chlorophylls is not known. In addition to the chlorophylls, chloroplasts contain carotenoid pigments and phycobilins. Carotenoid pigments are yellow or orange or red; they include the carotenes, which are polyene hydrocarbons (containing conjugated double bonds), and the xanthophylls which are oxygen derivatives of the carotenes.

All the carotenoid pigments are soluble in fats and fat solvents. The phycobilins have properties similar to those of the bile pigments, hence the name. They are water soluble compounds consisting of protein combined with a metal-free pigment group containing 4 pyrrole rings. Phycobilins are found in various algae. Both blue and red types are known. It is of course certain that chlorophyll is important in the photosynthetic process, but to what extent the other pigments play a role is not clear.

It should be emphasized that chlorophyll alone is powerless to invoke photosynthesis. It must be present in living cells. And yet isolated chloroplasts can accomplish at least part of the photosynthetic process. In the presence of light and in the presence also of ferric oxalate or of ferricyanide, the chloroplasts are capable of liberating oxygen. In the course of the reaction the ferric salt is changed to a ferrous salt. In place of the ferric salts, various other oxidation-reduction systems can be used, for example chromate or phenol-indophenol.

The oxygen liberated comes from water, and the hydrogen of the water acts as a reducing agent. But the hydrogen can not, by itself, reduce carbon dioxide, and the chloroplasts by themselves can not form carbohydrate. However if to the chloroplasts, enzymes of the cytoplasm are added, then carbon dioxide is reduced and a carboxyl group can be formed from it. This is an important step in the formation of carbohydrate, as will be shown later.

If the chloroplasts are broken up and transformed into a colloidal dispersion, some of the activity is retained, but if the chloroplasts are heated to 50° C., the activity is lost, indicating that an enzyme system is involved. In living cells chlorophyll is combined with protein. The chlorophyll-protein complex may be extracted from certain plants (lilies, beans) by grinding up the leaves and extracting in dilute sodium chloride solutions in the cold. The rate of photosynthesis increases with any increase in the amount of chlorophyll present in the plant cell, but the ratio is not constant at least in the alga Chlorella, for according to van Hille this ratio is highest in very young cells and decreases with age. In addition to carbon dioxide, light and chlorophyll, and the presence of unknown protoplasmic factors, the presence of water is of course essential to photosynthesis.

Temperature

There have been a number of investigations on the effect of temperature on the rate of photosynthesis. In general, both chemical reactions and biological processes proceed at a more rapid rate as the temperature is raised. Often there is a two- or three-fold increase with a 10° C. rise of temperature. However, photochemical reactions are known to have low temperature coefficients, that is to say, these reactions are not nearly so much affected by rise in temperature.

Warburg and others have found decided variations in the temperature coefficient of photosynthesis, and this fact has been regarded as indicating that the photosynthetic process involves two reactions, only one of which is photochemical. The general line of argument is summarized by Spoehr as follows: "At high illumination intensity a rise of 10 degrees from 15 degrees to 25 degrees causes the velocity of photosynthesis to double.

Here a chemical reaction is evidently determining the rate of the process; this reaction Warburg has designated as the 'Blackman reaction.' At low illumination intensity the rate of photosynthesis is independent of the temperature between 15 and 25 degrees. This is what would be expected if under these conditions a photochemical reaction, as, for instance, the photolysis of a chlorophyll-carbonic acid complex were determining the velocity."

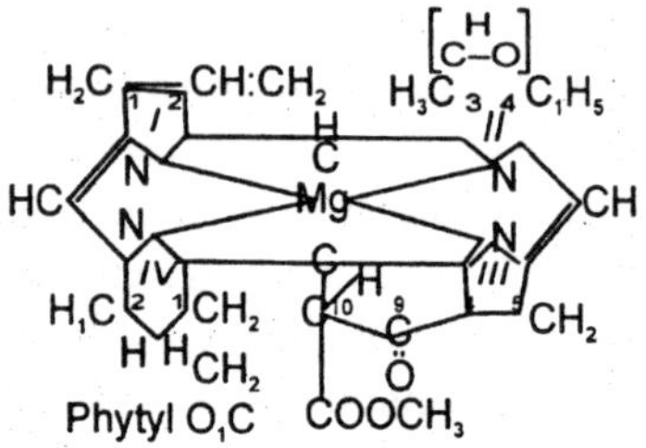

Under conditions of low illumination the photochemical reaction would be slow and, according to Blackman's principle of limiting factors, the slower photochemical reaction would then determine the speed of the entire process. Further evidence for a light and a dark reaction in photosynthesis is furnished by experiments with intermittent light which will be discussed later.

Other Factors. In the study of any biological process, a useful procedure is to discover the conditions which inhibit the process without destroying the life of the cells in which the process occurs. Many years ago, Claude Bernard found that fat solvent anesthetics inhibit photosynthesis without destroying the other vital properties of the green plant, and this observation has frequently been confirmed.

As will be noted in a later chapter, oxidative enzymes containing iron are inhibited by very low concentrations of cyanide, although as in the case of respiration in animal cells, the potency of the cyanide effect varies for different species. Very dilute solutions of hydroxylamine (NH_2OH) also inhibit photosynthesis. Oxygen in high concentration has a similar effect.

THE CHEMISTRY OF PHOTOSYNTHESIS

Because of the tremendous importance of photosynthesis—it is beyond much doubt the most important of all the chemical processes occurring on this earth—there has been a very great interest in the attempt to discover all that can be learned concerning the reaction, or rather the reactions, involved. This interest has been heightened by the fact that in animal cells some phase of the photosynthetic process may well occur. For it has repeatedly been shown that in animal cells, by the utilization of chemical rather than light energy, carbon dioxide and water can be synthesized into carbohydrate. In what follows, no attempt will be made to survey all the many-sided efforts that have been made to explore the chemical reactions that enter into the process of photosynthesis.

The barest of outlines is possible here. Fortunately there have been numerous reviews in which the subject is discussed in some detail. These have already been referred to. One important fact seems clear. In the general over-all reaction or series of reactions in the course of which carbon dioxide and water are combined to form carbohydrate, for each molecule of carbon dioxide used, one molecule of oxygen is released. This follows from a measurement of the volume of oxygen released and the volume of carbon dioxide taken up.

The ratio of these quantities, that is to say the volume of oxygen given off divided by the volume of carbon dioxide absorbed, is called the photosynthetic quotient and this ratio has consistently been found to be 1 or very nearly 1 by numerous investigators.

Accordingly, the over-all reaction of photosynthesis can be written as follows:

$$nCO_2 + nH_2O = nO_2 + (CH_2O)n.$$

But this equation tells us nothing about the intimate course of the reaction and that is what we want to know. *Induction Period*. When plants previously kept in the dark are illuminated, the photosynthetic rate is at first very slow and then increases until it reaches a constant level. The time interval before constancy is reached varies with the type of plant cells studied. Values of from two minutes to two hours have been obtained. The existence of an induction period of appreciable length is taken to indicate the necessity for the production of so-called "intermediate" substances which play a part in the photosynthetic reactions. In studying short induction periods, it is essential to use methods which are rapid enough to show quick changes in oxygen release or carbon dioxide uptake. McAlister employed a continuously recording spectroscopic method. Aufdemgarten and later van der Veen used a method which depends on the fact that the conductivity of carbon dioxide gas differs from that of air or of oxygen. This method also gave continuous recordings.

Dark and Light Reactions

It has already been noted that there is evidence to indicate that photosynthesis is the result of two reactions, one a truly photochemical reaction, and the other a reaction which may occur in the dark independently of light. The latter is called the "Blackman reaction." Warburg, studying the effect of intermittent light on the rate of photosynthesis of Chlorella, found that for the same amount of light, intermittent light was more efficient than constant light.

The effect was not so great with four flashes per minute, but with 8000 flashes per minute Warburg found a doubling in the rate of photosynthesis. Warburg used equal periods of light and dark. By making the light flashes much shorter than the dark periods, Emerson and Arnold were able to obtain a three- to four-fold increase in the rate of photosynthesis. These authors believe that the light reaction can take place in about 0.00001 second.

The dark reaception requires less than 0.04 second for completion at 25° C., and about 0.4 second at 1.1° C. The light reaction is dependent on carbon dioxide concentration and is inhibited by narcotics. The dark reaction is not noticeably inhibited by narcotics, is independent of carbon dioxide concentration, and is strongly inhibited by cyanide. However, Rieke and Gaffron believe that the reaction inhibited by cyanide is not closely related to photosynthesis itself. They think that cyanide prevents the initial reaction in which carbon dioxide is bound to organic substances within the cell.

Fluorescence Studies

The literature on chlorophyll fluorescence goes back many years. At present the field is a very active one; fortunately it has been very thoroughly reviewed. The early students of chlorophyll fluorescence recognized that chlorophyll in living leaves is much less fluorescent than chlorophyll dissolved in alcohol or some other organic solvent. In explaining this phenomenon, it has often been assumed that in leaves in which there is active photosynthesis, a portion of the light energy that is absorbed passes into chemical energy instead of being given off as light of another wave length. This is perhaps, to some extent at least, valid.

Various authors have studied chlorophyll fluorescence and photosynthesis simultaneously in the hope of learning something about the mechanism of photosynthesis. In the early studies, leaves were irradiated with ultraviolet light and the visible light given off was then measured. In these studies care had to be taken to avoid as much as possible the harmful effects of the ultraviolet. Later investigators have often used visible light to induce fluorescence.

The presence of oxygen tends to quench fluorescence of leaves. One of the leaders in the modern study of fluorescence has been Kautsky and on the basis of his experiments he has written out various hypothetical reactions in which chlorophyll is supposed to enter in the course of photosynthesis. Certainly one thing is clear: chlorophyll is necessary for the harnessing of light energy. But exactly how this harnessing is accomplished is not apparent, in spite of the extensive and careful studies that have been made of chlorophyll fluorescence. Wassink believes that in one way or another the dark reaction, involving as it undoubtedly does an oxidationreduction system of some sort, is converted by chlorophyll in the light to a system which can reduce carbon dioxide. Also according to Wassink, one of the enzymes that acts in the dark tends to be inactivated as soon as illumination strikes the plant and this inactivation is accompanied by a high degree of fluorescence.

When the enzyme is reactivated, probably as result of an oxidative process, there is a decrease in fluorescence. At present the subject of chlorophyll fluorescence and its relation to photosynthesis is too complex and too incompletely understood to throw any very clear light on the true nature of the photosynthetic reactions. Eventually, no doubt, as more work is done, the fluorescence studies together with the chemical studies that are so actively being prosecuted, will mean more than they now do, and the entire subject will be clarified.

Isotope Studies

Various isotopes have been used in the study of photosynthesis. These include heavy oxygen (O^{18}), heavy hydrogen (both H^2 and H^3), radioactive phosphorus (P^{32}) and radioactive carbon (C^{11} and C^{14}). Most of the early work was done with carbon 11. This isotope has a half life of only 21 minutes and as a result experimentation with it is somewhat difficult. Fortunately in 1945 carbon 14 became available. This is very long lived; its half life is of the order of magnitude of 10,000 years. The early workers in the field thought that the main facts concerning the chemistry of photosynthesis could be solved with relative ease, but as more and more work began to be done, the complexities of the problem came more and more to be realised.

When plants are exposed to carbon dioxide containing radioactive carbon, the radioactivity appears in various sugars and in various parts of the molecules of these sugars. It is true that the plant discriminates to some extent against carbon 14, and more sugar is formed from ordinary carbon which is carbon 12 than from carbon 14. But this is only a minor difficulty. In work with radioactive carbon, in order to discover intermediate compounds, it has been found advantageous to use very short periods of photosynthesis. The plant tissues are then destroyed and a search is made for the intermediate compounds.

One of the most successful methods that has been used in this search is paper chromatography, a technic very often employed by modern chemists. The most important intermediate compound thus far discovered is phosphoglyceric acid. In 1948, Calvin and Benson showed that this acid was the most abundant of all the radioactive compounds found after a short period of photosynthesis in the presence of carbon dioxide containing radioactive carbon. Since 1948, there has been much corroborative evidence. In one experiment, when green algae absorbing $C^{14}O_2$ had been exposed to light for 5 seconds, 87 per cent of the radioactivity was then found in phosphoglyceric acid. In an experiment with soybean leaves glyceric acid apparently appeared before the phosphoglyceric acid.

Triose phosphate seems to be formed from the phosphoglyceric acid—triose is a three carbon sugar—and hexose phosphates can readily be formed from the triose phosphates. In addition to these compounds, a number of other substances show some radioactivity soon after plants are exposed to radioactive carbon dioxide.

These compounds include malic acid, pyruvic acid and glycolic acid. The latter is a 2 carbon acid—it is simply hydroxy-acetic acid; and it may perhaps be a precursor of phosphoglyceric acid. An interesting type of work has been done with heavy oxygen (O^{18}). Using this isotope, Ruben, Randall, Kamen and Hyde and later Dole and Jenks were able to show that in photosynthesis the oxygen released comes from water rather than from carbon dioxide.

Quantum Efficiency

If a mol of glucose is burned in a calorimeter to carbon dioxide and water, approximately 675,000 calories of heat are liberated. It would require then the same amount of energy to build up the glucose from carbon dioxide and water. In photosynthesis, the energy involved in this process is derived from light. By determining the amount of light absorbed by photosynthetic cells, and determining at the same time from the oxygen given off and the carbon dioxide absorbed, the amount of photosynthesis occurring, it should be possible to know how much light energy is necessary to produce a mol of glucose; or on a molecular basis, how many units of light energy or quanta are necessary for the production of a single molecule of glucose.

If we divide the heat of combustion of a mole of glucose by the number of molecules in a mol (Avogadro's number), the quotient is approximately 3 times the energy carried by one quantum of light (at a wave length of 6,000 Å). Allowing for some loss of energy and considering also the fact that a single quantum of energy cannot be divided, it might be logical to assume that the minimum number of quanta required to reduce a molecule of carbon dioxide is 4 and that therefore the quantum efficiency is 0.25.

It might also be argued that the reduction of 1 molecule of carbon dioxide to a carbohydrate requires 4 hydrogen atoms. If in accord with the isotope studies on oxygen 18 we assume that the primary photochemical reaction is a decomposition of water, and we assume further that each hydrogen atom is contributed by a different molecule of water, then the necessary 4 hydrogen atoms would require for their liberation from water 4 quanta of light. Warburg was the first to investigate the quantum efficiency of photosynthesis and in an early paper as well as in later ones he did find a quantum efficiency of 0.25. Other workers for the most part failed to confirm this high efficiency, and found values of quantum efficiency of 0.1 or 0.08, indicating that 8-12 quanta of light are necessary to reduce a molecule of carbon dioxide to carbohydrate. The argument between those who favoured a higher and those who favoured a lower quantum efficiency has been rather intense, although the entire question may not be worthy of such ardent effort.

In measuring quantum efficiency, it is possible to determine accurately both the carbon dioxide taken up and the oxygen given off, and it is also possible to obtain a satisfactory measure of the light absorbed. One reason for uncertainty lies in the fact that during photosynthesis respiration occurs simultaneously and the magnitude of this respiration is not always clear.

Likewise, as noted previously illumination may cause an outburst of carbon dioxide which might obscure results obtained with a manometric method. Any warping of the photosynthetic quotient would change values of the quantum efficiency. Warburg himself now believes that in the presence of light there is a great increase in rate of respiration (in cells of the alga Chlorella). There is thus a back reaction, and this back reaction masks the true quantum efficiency. Indeed, Burk and Warburg find that only one quantum is required for the reduction of one molecule of carbon dioxide.

They believe that only a third of the energy required for the formation of carbohydrate from carbon dioxide and water comes from light. The other two thirds come from the chemical energy of respiration. This is a very interesting idea, for it brings closer together the processes of carbohydrate synthesis in animals and plants.

MODERN THEORY

Although there is much—very much—still to be learned about the chemistry of photosynthesis, some points now seem clear, and concerning these there is at the present time reasonably good agreement among the various expert workers in the field. In the first place, the action of light seems to be entirely, or at least very largely, on the reduction of water and the evolution of oxygen from water.

This reaction can indeed go on in the presence of isolated chloroplasts, light, and an oxidation-reduction system. But such isolated chloroplasts by themselves do not reduce carbon dioxide, and, as previously noted, other enzyme systems are necessary to accomplish this. All the reactions involved in the reduction of carbon dioxide and its conversion to carbohydrate may well occur in the dark. This is in accord with the fact that bacteria and animal cells, both of which are able to build up carbohydrate from carbon dioxide and water, can do this without any utilization of light energy.

In the course of the reactions which occur when carbon dioxide and water are synthesized by the plant to form carbohydrate, the intermediate compounds are to some extent at least the same compounds which appear when carbohydrate is broken down. Thus phosphoglyceric acid, which is the most abundant of all the substances found to occur in the few seconds after photosynthesis has begun, is well known as one of the compounds which appear during the course of carbohydrate fermentation.

Also, malic acid and pyruvic acid, which as already noted appear during the early stages of photosynthesis, are also important in the breakdown of carbohydrate. Moreover, some of the enzymes which play an important part in the formation of carbohydrate from carbon dioxide in animal cells may also be involved in the production of carbohydrate during photosynthesis. This is clearly indicated by the work of Vishniac and Ochoa, already referred to.

NITROGEN ASSIMILATION IN PLANTS

Green plants are able to synthesize amino acids and proteins from nitrates, nitrites or ammonium compounds. Such synthesis can take place in the dark, and it can apparently occur also in isolated roots grown in solutions of mineral salts, cane sugar and vitamin B1. There is thus no necessary relation to photosynthesis.

However, in wheat leaves there is some evidence that nitrate assimilation may in part at least be a photosynthetic process. There is still much doubt as to the exact nature of the various chemical reactions involved in nitrogen assimilation. In vitro, amino acids can be synthesized from keto acids (such as pyruvic acid) and ammonia, and it is probable that this type of reaction occurs both in plant and animal cells. Plant cells are able to reduce nitrates to nitrites, and then through intermediate stages to ammonia. Breakdown processes is the fact that cells not only break down carbohydrates to carbon dioxide and water, but can actually build up carbohydrate from carbon dioxide and water.

This was discovered only in relatively recent years. The direct evidence was obtained first for bacteria, then later for fungi, protozoa, and metazoan tissues. In rats, if carbonate containing radioactive carbon is injected, some of the glycogen in the liver can be found to contain radioactive carbon. The intermediate steps in what is commonly called "carbon dioxide fixation" have been much studied.

Buchanan and Hastings and also Wood have written reviews on carbon dioxide fixation as it occurs in heterotrophic organisms (that is to say, in organisms which require organic compounds and are therefore not autotrophic like the green plants and some bacteria). The enzymes which catalyze these reactions are the same enzymes long known to cause decarboxylation. In some cases manganous salts hasten the reactions. One of the most important enzymes involved is the so-called "malic enzyme," which is found in both plant and animal tissues; it can convert pyruvic acid and CO_2 to malic acid. Within cells generally, the synthesis of proteins is apparently carried on by the proteolytic enzymes which occur there.

Presumably these are the cathepsins. The subject of protein synthesis is discussed by Bergmann and Fruton and by Northrop. In order that a proteolytic enzyme promote the building up of a protein from amino acids rather than favour the reaction in the opposite direction, it is necessary that the end product of the reaction, the protein, be removed from solution, or that energy be applied to the system so as to change the equilibrium point of the system. Both mechanisms probably occur.

When the energy to shift the reaction is supplied by some other chemical reaction, the two reactions are said to be "coupled." There are various known examples of such coupled reactions. In protein synthesis, there is also a possibility that proteins may be produced from compounds other than amino acids. At any rate this is claimed by Alcock. Caspersson has noted that growing cells are rich in ribonucleic acid and he believes that ribonucleoproteins (pentosenucleoproteins) are primarily responsible for protein synthesis. According to Caspersson, the nucleolus and the region of the cytoplasm immediately surrounding the nucleus are especially important for protein synthesis. It is easy to make theories of this sort, but hard to prove or disprove them. However, Spiegelman and Kamen do present some indirect evidence that nucleoprotein does play an essential role in protein synthesis. The use of tracer elements has been of tremendous importance in the study of assimilation in animals.

One of the pioneers in this field was Schoenheimer, who died unfortunately at an early age. His book published posthumously, can be used as a guide to his researches, many of them done in collaboration with Rittenberg. One interesting point brought out by these authors is the speed at which assimilation occurs. Studies on fat metabolism of rats made with the aid of heavy hydrogen (deuterium) indicated that half of the fat present within the liver of a rat may have been synthesized within the preceding day. It takes somewhat longer for fats in the diet to find their way into storage depots. Schoenheimer estimates that the time necessary for half the storage fat to be formed is about a week. Proteins likewise are rapidly replaced. Thus the proteins and the fats contained in the cells of higher animals react with the diet at a rapid rate. There is a "metabolic pool" of substances from which an animal, an organ, or a cell obtains materials for the synthesis of its constituents.

The nitrogenous compounds of the metabolic pool constitute the "nitrogen pool." By feeding glycine containing nitrogen 15 and by determining the nitrogen 15 in the urine of rats and men, Sprinson and Rittenberg were able to estimate the rate of protein synthesis in these organisms. As is to be expected from its higher metabolic rate, the replacement of proteins in the rat is decidedly more rapid in the rat than it is in man. Shemin and Rittenberg had estimated the time required to replace half the proteins of the rat liver as 7 days. In some other organs and tissues the turnover of protein proceeds at a slower rate. This is especially true of muscle. For the rat, Sprinson and Rittenberg estimate the time required for the replacement of half the proteins of the whole body as 17 days; for man the time is much longer and is estimated as being 80 days. The rate of protein synthesis in the rat is 1 g. per kilo of body weight per day; for man it is only one fifth of this (0.2 g. per kilo per day).

The concept of rapid change in protoplasmic constituents originated with Schoenheimer and Rittenberg. It used to be thought that much of the food which entered the body served only as fuel and never really formed a part of the living substance. There was thus a distinction between endogenous and exogenous metabolism. The exogenous metabolism was thought to be concerned only with fuel substances, and this metabolism was sharply distinguished from the endogenous metabolism concerned with the protoplasm itself. This distinction is now shown to be false. To quote Schoenheimer: "The simile of the combustion engine pictured the steady flow of fuel into a fixed system, and the conversion of this fuel into waste products. The new results imply that not only the fuel but the structural materials are in a steady state of flux. The classical picture must thus be replaced by one which takes account of the dynamic state of body structure."

As our knowledge of assimilation processes increases, there is more and more indication of an interweaving of various types of metabolic reactions. As already noted, reactions which produce energy may in themselves serve as an aid to synthesis. Some of the complexities of the interrelationships of metabolic processes will be discussed in the chapter on intermediary metabolism.

GROWTH

In our discussion of metabolism up to this point we have followed a direct although, perhaps, winding path. We have considered, in turn, the types of food required by various organisms, the intake or ingestion of these foods into the body of an animal, the digestion of those foods which are solid and must be first changed into simpler, easily soluble compounds, the action of enzymes in digestion (and in other biological reactions), the absorption of dissolved food materials from the intestine of an animal to the cells of the tissues, the absorption of dissolved food substances by plant cells and, finally, the synthesis of relatively complex substances from the simpler food materials.

Arrived at this point, our path divides. It is possible, on the one hand, to consider further the building up of food materials into protoplasm and the resultant increase in the size of the organism. Or one might discuss the utilization of the food to produce energy: its oxidation. Such discussion would naturally involve the entire subject of respiration. Still a third possibility would be to discuss the transformations of the food material that intervene between assimilation and excretion. It is, perhaps, immaterial in which order these three topics are taken up. We shall proceed, therefore, to discuss them in the order named, concluding then with a discussion of excretion. Growth is a many-sided subject. The term itself is difficult to define, and biologists have on the whole been much more willing to study the phenomenon of growth than to attempt to define it in exact terms.

Thus, D'Arcy Thompson in his thoughtful book on *Growth and Form* states that "growth is a somewhat vague word for a complex matter, which may depend on various things, from simple imbibition of water to the complicated results of the chemistry of nutrition." Dictionary definitions are not much help; they usually define growth as "a gradual increase of a living organism." But when a man eats a meal he is not usually thought of as growing, nor is it generally believed that an elderly man grows when he takes on fat. Perhaps growth is best defined as lasting increase in volume during the period of progressive development.

As a matter of fact, in the practical measurement of growth, any increase either in volume or in length is regarded as growth. Philosophically, it might be wiser to define growth as an increase in the amount of protoplasm, as some authors do, but in actual experimental study such a definition would be impossible of application, for no one has established a criterion which enables him to distinguish the protoplasmic from the non-protoplasmic constituents of an organism. Under certain conditions, as for example starvation in flatworms or with increasing age in man, organisms can undergo decrease in size.

This may be called "negative growth." Child, however, prefers to use the term "reduction" instead of negative growth. Child's book should be consulted for a discussion of definitions of growth. There are various fascinating biological problems associated with the phenomenon of growth. It is because of differential growth that different animals and plants acquire their special forms, or that the shape of one organ or structure differs from another. Problems of this sort are considered both by D'Arcy Thompson and by Julian Huxley. There is, moreover, the question of the stimulus to growth, and the question as to why growth ceases. There is a definite limit of size for most organisms. A complete discussion of various growth problems would require many pages. We shall discuss, first, the chemical background of growth, and in particular the chemical substances necessary for and those provocative of it; secondly, the measurement of growth and, finally, the attainment of a definite limit of growth.

4

Communication System of Animals

Much of the work in ethology revolves around problems in animal communication. In fact, this is the work for which von Frisch is best known. He teased apart the "dance language" of bees. Worker bees dance to communicate the distance, direction, and quality of a feeding place they have found while foraging around the hive. The dance is performed on the vertical comb of the hive.

If the feeding place is directly towards the sun, the bees dance up. If the feeding place is directly away from the sun, they will dance down. As the sun moves across the sky, the bees adjust their dance so that the direction is always up-to-date. If, for example, a bee finds food at sunrise and indicates that it is located directly towards the sun by the time the sun sets, the bee will be dancing down. Thus, bees have an internal biological clock that tells them where the sun should be in the sky without them having to go out and check. Distance to the feeding place is communicated by the number of waggles at the center part of the dance. More waggles indicate greater distance.

Sites with lots of high quality food are communicated by the vigour of the dance (very good sites produce frantic dancing). In addition, other workers paying attention to the dancer touch their sister and can determine what type of flower she visited by the pollen stuck on her hairs. A second dance, a round dance, is used when the food source is very close to the hive. The type of communication an animal uses can be described by the sensory modality used to receive the signal. Examples can be seen in the following table:

Sensory Modality	Example
Chemoreception	Pheromones (group cohesion, territory marking, individual recognition, sex pheromones, etc.)
Mechanoreception	Vibrational cues including substrate-coupled (spider on web) and sound (bird and cricket song)
Visual	Bird courtship dances, aggressive stances of canids, firefly light patterns.
Electrical	Electric eels and weakly electric fish communicate with one another through charged pulses.
Psychic	That's how those voices in your head get there!

VISUAL COMMUNICATION

With good vision, it became possible for many animals to use a new channel for communication. Among those spiders that have good vision, courtship dances perfumed by male spiders help to get the female in the mood. Other animals have more spectacular, albeit still silly, courtship dances. Competition among males for the most impressive displays to attract females has led to ornaments, colours, badges, and other decorations to attract females and/or ward off other male competitors.

The red epaulets of the redwing blackbird, for example, are necessary for male birds to defend their territory. If the epaulets are painted black, the male can no longer chase off competing males (in fact, he is not taken at all seriously and is completely ignored by his rivals). Among other animals, the need to impress females has gone to an extreme. The male peacock, for example, is required to put up with his ungainly and brightly-coloured tail and accessory plumage to impress females. This is despite the fact that these feathers make him more susceptible to predation and are metabolically expensive to construct. The Irish elk is now extinct, and has been since the end of last ice age. During the ice age, competition among males and selection by females resulted in huge antlers used for display and defence.

When the glaciers retreated, forests grew in their place, and the huge antlers became a liability since the males couldn't move among the densely-packed trees without catching their heads.

This ultimately lead to the extinction of the species. Generally, sexual selection leads males to extremes and is driven entirely by female preferences and competition among males for mates. This has produced some extremely silly looking and conspicuous males over evolutionary history.

CHEMICAL COMMUNICATION

Pheromones are chemical "perfumes" that one animal emits to influence the behaviour of another. The pheromones can be classified according to the mode of reception (contact vs. airborne) or by their function (sex pheromones, aggregation, territorial, recognition, etc.). Both males and females lay out dragline as they move about. During their mating season (late summer or early fall, depending on the latitude), males find females by stumbling across their dragline. Pheromones deposited on the line alert the male that it was deposited by an adult female of his species and he'll follow it using his palps.

This is a left palp. At least two types of hairs are apparent on the palp. The more numerous pointed hairs are mechanoreceptors (a sense of touch for the spider). The brighter, less numerous, and curled hairs are chemoreceptors. Notice that the chemoreceptors. are found on the medial side of the palp (towards the upper left); that part of the palp used for following female's draglines.The curled hairs are the chemoreceptors, the straight ones are mechanoreceptors.

Note that both are inserted in sockets in the animal's integument to allow them to swivel when touched. Chemical communication has been found in nearly all organisms investigated, including fungi, bacteria, protists, and plants.

Among the animals, nearly all use pheromones to one level or another, with the notable exception of most birds. While animals such as spiders and insects tend to use pheromones with simple chemical structures, mammals use pheromones composed of a complex "cocktail" (with the exception of the social insects).

MECHANICAL COMMUNICATION

Communication through vibrations may be either substratum-coupled (through vibrations in the ground or other substrate), or acoustic (through the air or water). Among spiders, web vibrations are used to communicate between a courting male and the web's owner (he plucks out a song) or to determine the type and placement of prey that have blundered into the web.

To measure web vibrations in the lab, we use a laser system coupled to a computer. Vibrations of the web are picked up by a phototransistor whose output is sent to a computer through a digital-to-analog converter (DAC). As the web vibrates, different light intensities are reflected on the solar cell, producing a change in the voltage. This voltage change is assembled by the computer into a vibrational waveform.

The web vibration patterns differ among species. The image on the left is for *Pholcus opilionoides* while that on the right is for *Achaearanea tepidariorum.* Other animals that use a substratum-coupled vibration system for communication include water striders and leaf hoppers. Acoustic communication is accomplished as vibrations through air. Spiders also use sounds for communication. The oscillogram shows the pattern of pulses used by *Rabidosa rabida* males to communicate with females.

The lower spectrogram shows the timing of the pulse (along the X axis), the frequency (along the Y axis; lower frequencies at the bottom), and intensity (by the darkness of the traits; red is loudest). Compare the relatively simple sonogram from the spider to that of a redwing blackbird Note the richness of the spectrogram compared to that of the spiders.

IMPRINTING

A specialized type of "programmed learning" seen in many higher animals is a process known as "imprinting". The first experimental analysis of imprinting was done by Konrad Lorenz on geese hatchings. The first job of a gosling is to identify it's mom so they can follow her should she start to wander off.

Goslings are pre-programmed to follow the first object producing the species specific call. This object is then identified as "mom". It doesn't matter what the object is, just as long as it emits the proper call and moves. Researchers have had fun with this effect by imprinting ducklings on a model train that emitted quacks from the speaker. Ducklings would dutifully follow the train just as if it were mom. Another characteristic of imprinting is that this specialized learning must occur within a certain time-window known as the "critical period".

Ducklings that are first exposed to mom at 1 week of age will not follow here, since that is outside the critical period for learning the task. Song learning in some birds is entirely innate. Young doves raised without exposure to adult song will sing the adult song perfectly the first time they try it. Most other birds, on the other hand, need exposure to the adult song during a narrowly-defined critical period in order to imprint and learn the proper song. Swamp sparrows and song sparrows have a species-specific song.

The swamp sparrow song consists of a trill of nearly identical syllables delivered at a constant rate. Song sparrows produce a more complex song consisting of a trill followed by a melodious note phrase. Young swamp sparrows raised in the absence of adults sing a still produce a trill, but of less complex syllables (the innate song). Similarly, song sparrows produce an innate song that has some of the characteristics of the normal song.

Researchers have raised birds with and without the experience of hearing their species-specific song. Regardless of the chicks' experience, young birds all sing a variety of "call notes". Each of the symbols represents a different musical motif. Note that the experience of the animal doesn't affect this aspect of song development. If, however, a hatchling hears the adult song within the critical period (10-30 days in this example), they go through babbling and subsong stages differently so that, by the time they reach adulthood, their full song is very similar to the song they heard during the critical period.

In this example, there are slight changes in the song motifs and these song structures are similar to the original and differ only in timing, the range of notes used, or some other minor feature. These slight differences in song structure impart individuality to the songs, or dialects. The slight differences allow neighbouring birds to recognize one another as individuals. Birds stake out their territories by singing. After the territories are agreed upon, it is to everybody's advantage to recognize neighbours.

This way, a bird that hears a neighbour's call knows that he doesn't have to pay special attention to the situation since the territorial boundaries have already been agreed upon. Only when a stranger enters the territory and begins to sing does the male have to give it special attention (by chasing the interloper off). Note that some of the song motifs have a genetic component. The motif, for example, shows up in birds under both experimental conditions. Those with no song experience, however, sing a slightly different musical phrase vs. the structure for those with song experience).

If the chicks hear the song outside the critical period (before day 10 or after the first month), the lesson doesn't stick. These birds will sing as if they never heard the adult song. How does a young bird know which noises to pay attention to? Why doesn't a young Robin, for example, pick up the species song of a Blue Jay, or for that matter, a car horn?

The answer is that the young bird has some idea of what to listen for-an innate "neural template" that gives the general parameters of the species-specific song. The neural template theory suggests that the songs of unrelated birds do not fit the general configuration of genetically predisposed song structures. The template (or idea of how the species song should sound) is slightly sloppy; allowing for the creation of dialects and individual, bird-specific songs. Another example of restricted learning is food aversion. Animals that eat a novel food and then get sick will avoid that food from then on. In this example, a Blue Jay has eaten a Monarch butterfly. Monarch butterflies feed on milkweed as larvae and concentrate milkweed chemicals in their skin.

These chemicals make anyone who eats a monarch sick (at the very least; the toxins are cardiocides that attack the heart). Birds that have experienced the stomach upset brought on by eating Monarchs will avoid Monarchs from that day on.

OTHER AREAS OF RESEARCH IN ETHOLOGY

- *Comparative ethology:* The head-up posture is a widespread signal in fighting behaviour in many passerine families. The evolutionary explanation for this common behaviour is that they all share a common ancestor that performed this behaviour.
- *Analysis of Behaviour Patterns:* The sequence of activities during courtship of stickleback fish includes much aggression. The sequence of male courtship activities from dancing to fertilization are shown. The number of times that these behaviour patterns preceded or followed other actions is represented by the width of the connecting arrows. Attacks may follow any part of the sequence.
- *Play Behaviour:* Many animals play, not only as young, but also as adults. Play behaviour is thought to solidify social bonds, clarify rank within a social group, and prepares young for sexual and predatory behaviours. A common signal used to elicit play among animals is the butt-up display.

Ethology has moved into a technological direction. Ethologists are picking apart the neuro-endocrine basis of behaviour to explain the proximate reasons behind why animals act they way they do. There is also a movement to mechanize data collection so human error can be removed and strict statistical tests can be performed on raw data. Researchers have also begun to use molecular biology techniques to extract genes that control particular behaviours.

SEXUAL SELECTION

Sexual selection comes under 2 main classifications:

1. Competion can select for males with better fighting abilities, increased body size and weapons. Deers are an example with their large antlers for fighting – an example of sexual dimorphism.
2. Females making the choice as to which male they want as a mate.

If a change in trait increase reproductive success by improving chances of getting a mate then this change will be selected for, even if it decreases the overall probability of survival; it's all about getting the best genes possible into the next generation. This is why males take risks in mating and why juvenile males develop secondary sexual characteristics after maturation, to ensure that they don't suffer the costs of reproduction before they are ready for it.

INTRA -SEXUAL SELECTION

Intra-sexual selection is where one favours the ability of one sex to compete directly with one another for fertilizations. This is usually seen with male rivalry, and this situation is usually solved by fighting. Red deer (*Cervus elaphus)* ales grow antlers in autumn times to defend their harems from rival males during the mating season.

The females have no choice over who to mate with but they do gain because they are looked after by their male, if he can keep his harem then the females are likely to have strong male babies and have more reproductive success. When the males initiate the fight they start roaring at each other and the roars get louder and faster. If the challenger can not roar as loud or as fast as the other stag, then he would usually leave and not continue to pursue the harem.

However, this doesn't always work. In the event of this happening the rival males will size each other up, by walking around each other comparing their physical qualities with their own; for example, body size and size of antlers. The deers will try and not to fight with each other as the risk of getting hurt is high and the owner of the harem will not want to fight because if he gets injured he is likely to lose some females from his harem as opportunistic males will steal them. A deer will not fight if the chances of him losing are high, so if for example the challenging deer appears to have an advantage the defending one will back down and leave. This is an evolutionary high point (although it may not seem like it for the losing deer) as it means that the deers have become smarter and as a result live longer and in theory be able to produce more offspring, passing their genes for assessment on.

In many species the males look different to their female counter parts, they have a tendency to be more elaborately coloured. This is how they attract females to them and not other males. A good example and classic example of males looking more attractive than the females is peacocks, with their beautiful and big tail feathers. The male African Paradise birds (*Vidus paradisea)* have big beautifully coloured tails, but it is a handicap to them; because it is so big and heavy it gets caught easily.

Amo Zahavi said that it is a handicap to the males but because of it the males must be more tough and robust to be able to survive with the tails. Robert Fisher came up with a more plausible explanation in that if a female mates with an attractive male, she will have more attractive babies (males), which in turn will attract more females; evolution will end up running away for selection of more attractive males. A man called Andersson in 1982 did an experiment that very nicely showed Fischer's hypothesis; he set up 9 groups of 4 male long-tailed widow birds (*Euplectes prongue),* each matched for territory quality and tail length.

He took one male at random from each group and cut the tail to 14cm. the piece removed was then glued to another random male in the group. The other 2 acted as controls, but one had it's tail cut and stuck back on. He counted the number of nest in each territory; prior to the manipulation all males had on average 1.5 nests in their territory. After the manipulation, males with shortened tails averaged only 0.5 nests in their territory; males with lengthened tails averaged 2 nests in their territory. From this he decided that the females must favour the males with artificially lengthened tails.

RUNAWAY THEORY

Robert Fisher said that once selection for something starts, it ends up taking on a life of its own and produces mate choices for new reasons. He said that that an initial preference for males with a better chance of survival could end up evolving to the point where the males survival is reduced. He said that this process would exaggerate male ornaments until it was equalised with a reduction in male survival. An example would be bird tail length for example will continually get bigger if females prefer this and get to the point where practicality of tail length becomes insignificant. However, sometimes females' preference is so strong that despite a reduction in male fitness, the population can end up becoming extinct. It is self-reinforcing as every time a mutation arises for a particular trait and is a favourable one, it spreads through the population. He said that this type of behaviour is beneficial to choosy females as their offspring would then benefit. Her daughters would prefer more attractive males, as her genes for sexual preference will be passed on; and her sons would be more attractive as they gained their fathers genes. These so called favourable genes are continually added to the gene pool

HANDICAP THEORY

How can a female realistically choose "good" genes when in the wild all she has to go on is looks? However Amo Zahavi suggested that certain physical attributes, like ornate tails, showed off a male's hardiness. That they can survive despite having this handicap and therefore must be more fit. He said that if a female chose a male with a long tail for example is guaranteed to be a high quality mate because only high quality males can afford to be burdened with a long tail. Hamilton and Zuk extended this idea 1984 by suggesting that the males must be healthier because they are able to "advertise" that they are free of disease and parasites.

As disease is a major source of juvenile mortality this may have been the driving force behind this evolution of sexual preferences. If the male is free of disease, he must have better genes to combat disease and if the females mates with him then her offspring are likely to also have these genes. Also, if a male has a beautifully kept tail, for example, then this demonstrates his potential parental care in looking after the young.

OPERANT LEARNING/ASSOCIATIVE LEARNING

Animals can learn to associate a voluntary action with something that happens. This type of learning is done by trial and error; B.F Skinner made "skinner" boxes in which he placed rats. After a while of these rats being in the box they would accidentally stumble upon and knock a bar which would release food into a food hopper, now, the rats wouldn't notice immediately that the food had been released, but would fairly soon after pressing the bar.

This process of the rat knocking the bar and finding the food would happen a number of times (as the rat is trying to find a means of escaping) until they realise that the food is appearing because of the bar pressing and they will then continually press the bar over and over again to get more food out. The rats had learnt that it would be rewarded with food it they pressed the bar. Another really famous example is Ivan Pavlov and his dogs. He would restrain a hungry dog in a harness and would give it small amounts of food. Whilst he did this he would regularly ring a bell.

He wanted to measure the effects this would have of the dog and to see if it learnt that it would be fed when the bell rang. He measured the amount of saliva that the dog produced. He found that after a while of the dog associating the bell with being given food the dog would actually salivate more and more. The bell is known as the conditional stimulus and the food was the unconditional stimulus; so when the dog salivated over the food, we were seeing an unconditional response as this was something that the dog would do regardless but the salivation when the bell rang was a conditional response.

The dog salivated when the bell rang because it knew that it was about to be fed, it got to the point where you could ring the bell after and during being fed that the dog would salivate more profusely. Animals in the wild learn exactly in the same manner, for example, if they see some food that they haven't come across before, eat and find that it tastes nasty they will not eat it again.

They will then remember what it was and associate the colouring and patterning with other foods and know not to eat these either. For example some birds will learn after a few tries that the black and yellow cinnabar moths taste evil and then not eat anything that is black and yellow, like wasps.

INFANTICIDE

Infanticide is the culling of juveniles that aren't sired by the new male that has taken over a group or pride. This type of behaviour is shown in many species of animals, but is especially seen in lions. But why it is advantageous for a lion to kill cubs that aren't his when he takes over the pride? The females could quite easily turn on him if he attacks their offspring. It doesn't seem likely that this behaviour has evolved by natural selection, it could possibly have evolved by means of social pathology in order to prevent over crowding. V. C. Wynne-Edwards says that this happens for the good of the group, otherwise the group would become over crowded and then eventually the extinction of the group would occur due to all the critical resources being used.

He argued that this lead to evolutionary change among different groups and the existence of self sacrificing individuals. With the multiple allele idea it was easier to see that there were selective pressures acting on individual in a group, which would have more evolutionary impact. However, this is not such a convincing idea. Sarah Hrdy believes that it is quite possibly due to sexual competition between males. As it might increase the lion's reproductive success because if the cubs are dead then the females will want to mate again; so he gains descendants and his genes therefore enter the gene pool. This is a more plausible idea because if the new male does not kill the cubs that are under 9 months old then it will take 2yrs before he is able to mate with the lionesses. If they lost their babies due to infanticide it was found that the females conceived within 4.5months of the new male taking over the pride.

This sort of behaviour should only happen with abnormal/ unusually high population sizes. Unfortunately this is not the case, as seen with langur populations that are untouched by humans and with moderate population sizes. So which theory is correct? Sexual competition is definitely part of it. But is cannibalism? (Male lions kill cubs up to 9 months old) They can replenish energy lost after ousting the old male, but in Langur communities this act of cannibalism is not seen when new males just simple join a group. It is most likely that the new male is showing a form of selfishness and kills the young so that he benefits by getting his genes into the gene pool quicker.

ALTRUISM

Altruism is behaviour that benefits others at a cost to one's self. Song birds display altruistic behaviour as they feel obliged to sit on eggs that are present in their nest. Cuckoos have exploited the song birds' good parenting and lay their eggs in the nests. Once the cuckoos have hatched they eject the song bird eggs and babies from the nest.

However, Lotem *et al* had noticed that song birds have picked up upon this parasitism and have learnt to recognise their species markings on the eggs so that when they spot a different species egg in their nest, they eject the egg before the cuckoo can do any damage. But even the cuckoos have got smart and noticed that their offspring weren't surviving and started to make their eggs look more like song birds eggs. Richard Dawkins described it as an evolutionary arms race in his book *The selfish gene*.

He noticed that for cuckoo's there was selection in favour of eggs that look like song birds; as these are the ones that make it into the next generation and pass on their egg shell design genes. Also there was selection for song birds with sharp eyesight, as it was these birds that spotted the cuckoo egg wasn't theirs. So if their eggs then survive, then so do the sharp eyesight genes.

What is also unfortunate for the song birds is that if it is their first brood and the cuckoo egg does hatch, when they have a second brood, they are most likely to kill their babies as the parents are expecting baby cuckoos.

GROUP LIVING

Meerkats *(Suricata suricata)* are very social animals, and each animal in the group has a role to do in order to secure the safety of their offspring. A select few in the group operate as "look outs" to inform the group when a predator is coming. They sit up in trees and keep a vigil. From an outsiders' perspective it would appear that the "Look outs" show altruistic behaviour; for when they are on guard they do not feed. However, they live in a very close nit society and they benefit enormously by doing this. Other members in the community are aware of who the individuals are that act as "look outs" and provide them with food, etc. Group living has strong advantages, Lima proposed his Model of Co-operative Vigilance; he described this as a trade off between foraging and vigilance. The more vigilant an individual is the more increased is each individual's survival due to the dilution effect.

VIGILANCE

It's important for an animal to keep a vigil, especially when eating food, for you never know what might eat you when you least expect it. With some birds it is known that if they feed on their own they spend a more time looking up and about than when they are in a group. They can share the responsibility between them. G.V.N Powell took some starlings and put them in an aviary. He kept some on their own and others in group sizes of about 10. He made an artificial hawk that flew over them and he noticed that birds on their own to took a lot longer to react than the birds in groups. So, even though it may be in some aspects better to forage on their own to reduce conspicuousness for example it is better to work as group and take in turns at looking out for predators as they will be able to flee quicker in danger. But this can mean that some animals will cheat and not do their part in looking out for the group.

DILUTION EFFECT

The dilution effect is simply that the more individuals there are in a group, the smaller each individual's probability of getting attacked is. An example of this is seen with Adélie Penguins. They are predated by Leopard Seals, which loiter around the edges of the ice formations where these penguins are found, so that when the penguins dive in the water to feed they ambush them. So, if the penguins wait until there is a big group of them, then they can all jump in together, reducing the risk of getting caught. Unfortunately, those that jump in first or last are most likely to get eaten. Will Cresswell in 1994 observed and carried out experiments on Redshanks in mudflats and marshes in Scotland. He found that once flock sizes reach over 30 in number that each individual bird spent less than 2 seconds at a time being vigilant. He therefore realised that once communities reach a certain size, vigilance no longer has such a crucial effect on the community. He noticed that it actually became harder for an individual to be singled out by a predator for attack and some times it would even deter a predator. This is another good example of the dilution effect being seen in practice.

CHOOSY FEMALES

The main problem for females is to choose a male as a sexual partner that will ensure her babies will have the greatest chance of survival and reproduction. As a result of this one would assume that a female would need to be picky when it comes to choosing a mate. This is not always the case as some males will take a route of forced mating with female, but this is not a very successful form of mating as usually the female will have to up a certain position to be receptive for successful reproduction.

Being choosy isn't always of a benefit to the females and as a result she may actually suffer a reduction in fitness; you expect that when experience direct selection the females' preference will evolve to an optimum that would maximise fecundity. To reduce the risk of decrease their fecundity females are coy to gauge directly the male's fitness. George Williams described courtship as a contest between male salesmanship and female sales resistance, there has to be a compromise at some point. The females will prolong courtship to find out what the male is really like, his true nature. Natural selection favours distinct features, by prolong courtship, the female might be able to see some of his distinct features that could her aid her children's' survival.

In Herring gulls courtship (*Carus argentatus*) the female will beg like a juvenile does to be fed so that the male can feed her. However, the passing of food doesnt always take place in some species of gull; its just a ritual. With the common tern (*Sterna hirudunda*) Nisbet thought that this form of courtship effects the clutch weight and is thought to be a predictor of the males performance in feeding their young, he felt that it was a good indicator of his parental responsibilities.

LEKS

Leks are display sites in which males congregate and females visit only to find a mate. There are theories as to why lekking occurs. The males have small territories that usually lack useful resources for the visiting females. After reproduction occurs the females leave and most times never see the man again. The number of matings that each male will have in the lek is by no means equal. When populations come under direct selection, you would expect female preference to evolve to a maximum fecundity; leks can sometimes be seen as a bit of a paradox because female choice of a mate does not influence her fecundity but her mating preferences have evolved under the circumstances. However, recent studies have shown that this may not be the case. Robertson and Bourne both did independent work on the frogs *Uperolia laevigata* and *Ololygon rubra* (respectively) and found that the females would select males of certain sizes to maximise fertility probability due to the difference in size between the 2 sexes affecting the actual mechanism of external reproduction. Jack Bradbury says that for groups that use lekking for reproductive means do this because no other method is as good. It sometimes happen because females will often travel many miles to feed, so if the men are crafty and wait in the trees where she'll get her food, they stand to get a better chance of mating. On the other hand, so males will group in areas where there is absolutely no gain for the females, the females then have to actively search for a mate.

Three theories have been proposed as to why lekking occurs:

1. *Female hotspots*: Males gather where females tend to travel
2. *Male hotshots*: Ugly males that are less likely to get a mate on their own hang around good looking males in the chance of getting noticed.
3. *Female preference*: The males cluster because the ladies like to be able to have a bigger choice of who they want to mate with.

These 3 theories are all seen in practice in the real world, it all depends on how the population has evolved to which theory fits best.

HOW DO FEMALES CHOOSE

Firstly, how do the females gain from leks when they only get genes for their children. Theorists argue that females will only have a limited choice in genes because genetic differences in males would end up being phased out if the females just pick the most attractive males in the lek. However, direct measurements have been taken on genetic variation and it shows that a high proportion of variation in male sexual characteristics comes from genetic variation (beauty is after all in the eye of the beholder); therefore females still do have a wide choice of characteristics.

When in a lek, choice is primarily on genes as there is no parental control.

There are 3 ideas on how a females makes the choice:

1. They prefer males who display honest costly signals that they have superior genes – this idea is based on the handicap theory.
2. Females prefer males with higher fitness and their genes will come together making sure her preference genes get into the next generation, they work on traits for males that may be disadvantageous for them – this idea is based on the runaway theory.
3. Females may actually inherit sensory capacities from their ancestors that bias the traits that they select for in males. For example colour blind females will select males with striking black and white patterns

These indicator mechanisms means that females can select males that can sire offspring with superior fitness

ADAPTIVE EVOLUTION

Daan *et al* in 1990 took eggs from kestrel nests and put them in other kestrel nests to increase clutch size artificially. He found that by doing this once, the subsequent clutch survival increase to 70 per cent after the 3 rd year. He found that with enlarged clutches, more fledged and that their overall reproductive value had increased. However, even though the chicks in the larger clutches had increased fitness, their parents had actually suffered a decrease in fitness as they ended up having more responsibilities and less time to ensure their own survival. This enlarged clutch size in the end decreased the residual reproductive value of the kestrels, therefore decreasing clutch size. So, the size at which the clutches were originally before Daan had tampered with them had evolved to be that size by means of natural selection as a compromise between fitness and fecundity.

5

Behavioural Studies in Zoo Conservation

Much early and influential research in animal behaviour and comparative ethology has been carried out on captive-held animals and in zoo settings. Indeed, the development of ethology as a scientific discipline and the beginnings of zoo behavioural research complemented each other and occurred concurrently in the mid-1940s. The work of Hediger and Meyer-Holzapfel in the early 1950s pioneered behavioural zoo research in Europe.

In the early 1960s, Desmond Morris started the first formal zoo behavioural research unit at the Zoological Society of London, Regent's Park. Soon thereafter, the New York Zoological Society, in association with the Rockefeller University, formed the Institute for Research in Animal Behaviour, and in 1965 the National Zoological Park founded its research unit. Behavioural studies have predominated in zoos for decades.

Animal behaviour studies have contributed significantly to the improvement of management and reproduction of zoo animals and have been regarded as a powerful tool for captive management. However, many behavioural research papers, although important in their findings, are the result of casual observation or purely descriptive case studies.

They frequently arise from an effort to solve a particular crisis encountered with captive-held individuals or species. Solutions are applied over time until breeding success is obtained, and biological understanding of the obtained solution may or may not be achieved during this process. The scientific method using systematic hypotheses testing has rarely been employed in such studies. Recent technological advances in wildlife immobilization and anesthesiology, biotechnology, genetics, and reproductive physiology have received increasing attention in the search for solutions to breeding problems.

The development of techniques for assisted reproduction (*e.g.*, artificial insemination, *in vitro* fertilization, and embryo transfer) has contributed much valuable information to our overall understanding of the reproductive biology and physiology of many species. Furthermore, in some cases the use of assisted reproduction may provide the only feasible way to save a species or subspecies from certain extinction. Theoretically, techniques for assisted reproduction can offer a way to increase species capacity and to facilitate the movement of animals without the hazards of disease, behavioural stress, and social problems.

Similarly, importing genetic material from wild into captive populations can be accomplished without further depletion of wild specimens. However, so far these technologies have only been worked out for less than three dozen wild species of mammals and birds, and broader application may not be feasible due to limited resources. Furthermore, for species in which such technologies are available, concurrent behavioural studies are urgently needed to examine possible drawbacks of the use of assisted reproduction. Shepardson appropriately cautioned that there is more to preserving species than looking after their genes. The intended goal of captive-breeding Programmes is to preserve animals as representatives of their wild ancestors for posterity, research, education, and for potential release back to the wild.

Thus, preservation of natural behaviours should be a key factor for all these endeavors. Failure to reproduce naturally and raise young in the captive environment will most likely result in a loss of many forms and patterns of natural behaviour. Many mammalian species show a large repertoire of complex learned and culturally transmitted behaviours (*e.g.*, sexual and parental competence, social interactions, some aspects of foraging behaviours). Such behaviours may be lost permanently if no provisions are made to support their development and display in captivity.

Indeed, the diversity of culturally transmitted behaviours has the potential to be lost much faster than genetic diversity. It is therefore of utmost importance that captive animals develop behavioural competence in all areas important for living in the wild if the ultimate goal is reintroduction. Provision of more naturalistic environments (*i.e.*, simulating a species' social and physical environment found in the wild) may help to minimize the effects of domestication or selection for a specific captive environment, and successful unassisted captive propagation of a species may serve as an indicator of our intuitive or scientific understanding of a particular species' requirements. In short, systematic research efforts are urgently needed to examine various aspects of behaviour pertinent to conservation efforts for captive and wild populations of rare and endangered species.

Below, several areas of behavioural research that are particularly relevant for captive-breeding success in mammals are identified. Examples of successes and failures of captive propagation that can be attributed to behavioural factors and that illustrate the importance of behavioural research are presented in each section.

ASPECTS OF ANIMAL BEHAVIOUR

TERRITORIALITY AND SPATIO-STRUCTURAL REQUIREMENTS

Hediger first noted the importance of territoriality and functional space for captive management of wild animals. Knowledge of species-specific movement patterns and territory requirements is essential for maintaining close-to-natural social and reproductive interactions in mammals. In addition, perceived space, including olfactory and auditory components, should be considered together with specific niche requirements for successful maintenance and breeding of captive mammals. Providing appropriate space objects (*e.g.*, climbing structures, hiding places) has been shown to have substantial behavioural effects by reducing aggression and stress in several species.

For example, in a pair of giant pandas *Ailuropoda melanoleuca,* a significant decrease in frequency of agonistic vocalizations occurred after addition of a complex series of platforms, swinging tires, and other mobile features to the enclosure. In leopard cats *Felis bengalensis,* reduced exploratory behaviour was an indicator of chronically elevated adrenocortical activity (indicating stress), and hiding behaviour increased with elevated adrenocortical output.

Consequently, provision of appropriate hiding places stimulated exploratory behaviour and reduced pacing and urinary cortisol levels. These results indicate that structural enrichment of enclosures may increase behavioural opportunities for coping with aversive stimuli. Cage size has been shown to affect behaviour and reproduction when crowding occurs. Groups of rodents and nonhuman primates in captivity exhibited higher frequencies of aggressive behaviour than less crowded or free-living conspecifics. Blake and Gillett found that an increase in cage size resulted in larger litter sizes and increased overall breeding success in Asian chipmunks Tamias sibiricus.

In other rodents, cannibalism and delayed implantation have been observed as a result of crowding. Even when cage size is appropriate, sounds and odours from adjacent enclosures may lead to a perceived crowding effect or territory invasion reported that in an off-exhibit breeding facility for golden lion tamarins *Leontopithecus rosalia* where an air-handling system inadvertently also moved odours and sounds from enclosure to enclosure, neighbouring family groups frequently seemed agitated and vocalized in a distressed manner. Among primates an inappropriate captive environment is frequently associated with stereotyped movement, self-aggression, higher levels of aggression between cage mates, inactivity, and infertility. Indeed, numerous studies on a variety of mammal species have documented an increase in natural behaviour patterns and a decline is stereotypic movements after providing various enrichment devices and increasing the complexity of the captive environment. In some cases reproduction may be directly and positively affected by such changes. For example, several previously nonreproductive gorillas (*Gorilla gorilla*) became reproductively active after transfer to an enriched enclosure. For some species certain enclosure features may be necessary simply to increase survivorship in captivity.

For example, in several ungulate species fatalities have been reported due to inadequate enclosures or enclosure size. Death resulted from collision with fences or other enclosure structures.

In pronghorn *Antilocapra americana* the importance of distant-visibility has been emphasized, and recommendations for fencing and shelter constructions have been made to accommodate natural behaviours and reduce fatalities. Furthermore, species-specific enclosure design needs to consider appropriate hiding places, denning sites, substrate, and other structural features. For example, the giant panda spends little time auto-grooming but requires soil or grass and water to allow for regular rubbing and bathing to maintain a healthy fur condition.

INTERSPECIFIC INTERACTIONS

Some species are maintained in mixed-species exhibits. This practice is particularly common in ungulates. However, excessive aggression and even fatalities have sometimes occurred as a result. Zebras (*Equus* spp.), for example, have been observed to kill wildebeest (*Connochaetes* spp.) and giraffe calves (*Giraffa* spp.) at some facilities. In a multi-institutional study on interspecific aggression in 10 ungulate species, a correlation between aggressive behaviour and taxonomic relatedness was found.

Interestingly, aggression was found to be highest between distantly related species. A possible explanation offered for this phenomenon was the inability of distantly related species to recognize each others' threat displays and thus avoid further aggressive interaction. However, there are encouraging accounts of various attempts to establish mixed-species exhibits, and some have reported an increase in the behavioural repertoire shown by species under mixed-housing conditions. For small to medium-sized predators, such as cheetahs *Acinonyx jubatus* and a number of small felid species, it has been reported that housing next to large-sized predators and natural competitors, such as lions *Panthera leo* and leopards *Panthera pardus,* may decrease reproductive success. Although recently there has been increased awareness that some species may not thrive well in captivity if in constant, close proximity to their natural predators there have been few systematic studies to investigate this contention.

One study indicates that ungulates may change their behaviour as a result of constant visual presence of predators. However, the authors concluded that because the changes only occurred in rare behaviours, juxtaposing predators and their prey in exhibits did not result in serious behavioural changes.

FORAGING BEHAVIOUR

Wild mammals usually spend a significant amount of time foraging. For instance, wild orangutans *Pongo pygmaeus* spend about 46 per cent of their time feeding, wild gorillas about 45 per cent and chimpanzees *Pan troglodytes* about 74 per cent. During foraging wild animals face decisions about where, when, and how long to forage, and which items to choose. However, most captive mammals are deprived of such experiences and are frequently provided with uniform diets presented at the same time and place every day.

At one facility foraging time in captive gorillas was reported to have increased from 11 per cent to 27 per cent after the provision of browse; another study on captive gorillas reported about 8 per cent of time spent on feeding and foraging.

It is likely that severely reduced levels of foraging behaviour are prevalent for most captive mammals when compared with their wild counterparts, and variation of food presentation over space and time may allow increased foraging activity in captive specimens. Foraging may be associated with developmental, parental, and sexual behaviour for many species. Provision of more varied diets and the possibility to exhibit natural foraging behaviour may therefore be important for breeding success.

Breeding success may also be influenced indirectly by enhancing the general well-being of the animal or by directly stimulating sexual activity. For example, exhibition of more complex feeding behaviours decreases stereotypic behaviours and increases general activity and well-being in different species such as macaques, orangutan several other primate species, and leopards. In the cheetah the stimulation provided by catching and killing live prey appears to enhance breeding success in some instances. Hediger noted that feeding wild animals in captivity should meet nutritional requirements but also needs to fulfill psychological and behavioural requirements.

Detailed knowledge about natural foraging behaviour and foraging patterns is necessary to provide appropriate variety and appropriate temporal and spatial presentation of food items. Aggression in a group of stump-tailed macaques *Macaca arctoides* was reduced by presenting food scattered on to a wood-chip litter substrate.

Similar results were obtained for pig-tailed macaques *Macaca nemestris* and for chimpanzees. In the light of preserving natural behaviours for possible reintroduction, the importance of learning species-specific foraging patterns may be crucial for success. For example, adult dietary preferences of black-footed ferrets were found to be established at an early developmental stage during the third month of life.

SOCIAL ORGANIZATION: PATTERNS OF GROUP FORMATION AND DISPERSAL

The importance of providing an appropriate social environment for successful captive propagation has frequently been emphasized. Nevertheless, many species are still held in arrangements that consist of forced social living of at least one adult of each gender being permanently housed together. Species that require different social conditions may show little to no reproductive success or will be unable to rear viable offspring as a consequence. Our understanding of the social interactions of endangered species is frequently based on anecdotal information or few rigorous studies. For example, most felids are simply regarded as asocial, yet data from field studies only exist for a small number of felid species.

In one case, extensive field studies revealed an unexpected pattern of sociality given the species' solitary nature: in the cheetah, males formed long-term stable bonds termed "male coalitions". Thus, patterns of sociality even in largely solitary species may be more intricate than assumed from minimal observations or extrapolation of findings in related species. Furthermore, extrapolation from closely related species may frequently prove inadequate. For example, three zebra species, Burchell *zebra Equus burchelli*, Grevy *zebra Equus grevyi*, and mountain zebra *Equus zebra*, were found to behave quite differently from each other with regard to herd formation. False assumptions about the social nature of an endangered species, based on extrapolations and scant field data, are not uncommon, and detection of such errors is critical for long-term captive-breeding success.

For example, the aye-aye *Daubentonia madagascariensis* was originally described as solitary, but later findings indicated a more social nature. Matschie's tree kangaroos *Dendrolagus matschiei* were traditionally housed in pairs or small groups, but behavioural studies on captive and wild tree kangaroos suggest a largely solitary existence. The brown hyena *Hyaena brunnea*, although belonging to its own family (*Hyaenidae*), had mostly been kept in pairs, a social structure successful for many canid species; however, field studies showed that this species is far more social than had been assumed.

Elephant shrews *Elephantulus rufescens* were often kept in groups but never reproduced in this situation. Field research indicated that most elephant shrew species might be monogamous. Once housing arrangements in captivity had been changed and elephant shrews were kept in pairs rather than in groups, successful reproduction occurred. Similarly, red pandas *Ailurus fulgens* were found to be unable to rear offspring when two or more females were housed together. Among marmoset and tamarin species, founding pairs were often allowed to develop into large family groups. The offspring, however, would never reproduce in these family settings.

Periodic removal of adolescent young and their establishment as new pairs resulted in release of reproductive suppression and was therefore crucial for a successful breeding Programme. Various types of reproductive suppression have been documented for a number of species, where either offspring remained within family groupings past the time of sexual maturity—*e.g.*, bush dog *Speothos venaticus*, cottontop tamarin *Saguinus oedipus*, naked mole rat *Heterocephalus glaber* or where individuals with low dominance status were forced to remain in inappropriate social settings—*e.g.*, black-howler monkey *Aloutta caraya* cotton-top tamarin Japanese macaques *Macaca fuscata*, marmosets and tamarins.

This may be part of a reproductive strategy for some species, where alloparental "helpers" are necessary to rear offspring successfully in the wild. However, for other species it may be a reflection of inappropriate social conditions in captivity. Therefore, detailed knowledge of natural patterns of dispersal and grouping is necessary to determine species-specific management decisions. Inappropriate social composition has also been observed to lead to abortion or unsuccessful reproduction. Furthermore, introductions of unfamiliar individuals into established groups without knowledge of dispersal patterns in the wild may lead to serious injuries, unexpected fatalities, or death of the newcomer. For some ungulates, the establishment of stable female hierarchies has proven crucial for reduced aggression and successful reproduction.

Furthermore, there is evidence that a minimum group size may be necessary for successful reproduction in some species. Such social facilitation of reproduction, also termed the Allee effect has been documented for several bird species. However, it may also occur in mammals. For example, in black lemurs *Eulemur macaco macaco* and mongoose lemurs *Eulemur mongoz*, reproductive success may be enhanced by the presence of additional conspecific pairs. In Przewalski's horse, white rhinoceroses *Ceratotherium simum*, and vicunas *Vicugna vicugna* larger groups have been found to dissipate aggression of males and avert fatalities in females, and they may also act as a stimulus necessary for breeding.

MATING SYSTEMS AND REPRODUCTIVE BEHAVIOUR

Consideration of species' mating systems is one of the most basic requirements of captive propagation. Historically, however, it has been common practice in zoos to house only a single pair, male and female, of each species. Mating systems in mammals are highly diverse; 11 different mating systems have been described in one classification and only about 3 per cent of mammals have been classified as monogamous.

Thus, monogamous pairings were only successful for a small number of species and individuals. In many species the natural mating system may not be obvious from the social composition observed in the wild. For example, in packs of wolves (*Canis* spp.) and African wild dogs, only one pair will form a monogamous unit and reproduce; however, to a casual observer, polygyny, polyandry, or promiscuity may be inferred based on social composition alone. Even when the general mating system is known (*e.g.*, giant panda, cheetah), our understanding of the underlying behavioural-physiological mechanisms of mate location, mate choice, timing and induction of sexual arousal, and courtship behaviours may be insufficient to create the appropriate environment for successful reproduction.

This is particularly apparent in asocial species, where mating represents an unusual event of "sociality" during which the tendencies to avoid conspecifics are overcome for a short time. Mating in these species may involve a substantial risk, and aggression is usually part of the courtship ritual. There are numerous accounts of captive animals where mates have killed or severely injured each other. Behavioural studies have offered successful solutions in some cases. For example, Thompson's study of captive kangaroo rats suggested that individuals destined for breeding should be kept within sensory contact of the opposite sex to allow for socialization and familiarization. Long-term familiarity appears to mitigate aggression among males and females. Similarly, in clouded leopards and leopards, one strategy of achieving compatible breeding pairs is to place sexually immature individuals together for bonding and familiarization. However, long-term familiarity of breeding partners can also be detrimental to breeding success, particularly in species where avoidance of mating with closely familiar individuals may be part of a strategy to avoid inbreeding or to facilitate dispersal.

Introductions of an unfamiliar estrous female can lead to sudden sexual interest in previously sexually unresponsive males. For the cheetah, an important provision for repeated successful breeding is to isolate females and only occasionally bring males and females together for breeding introductions. More generally, the presentation of novel stimuli such as live prey or an altered environment may also stimulate reproductive activity in some species (*e.g.*, cheetah, red panda, white rhinoceros).

Courtship chases or courtship-associated fighting may also be a necessary part of reproduction (*e.g.*, cheetah: black rhinoceros). For several gregarious ungulate species (*e.g.*, white rhinoceros, Asiatic wild ass *Equus hemionus*, Przewalski horse), it has been reported that keeping males housed singly or only with one female may result in severe aggression and even death upon reintroduction to a herd. Understanding mechanisms of mate choice and patterns of individual variation is also crucial for successful breeding of many species.

For example, a survey on random pairings of tree shrews revealed that only 20 per cent resulted in successful pair formation. Mate incompatibility and rejection of apparently reproductively healthy individuals presents a major problem for breeding facilities housing small numbers of individuals of a particular species. Mating recommendations in captive-breeding Programmes based solely on genetic and demographic considerations may therefore conflict with natural mating strategies and may lead to unproductive management decisions.

PARENTAL CARE

Management practices that provide the appropriate environmental and social requirements for natural parental behaviour are critical for successful propagation of captive mammals. In particular, the survivorship of young may be affected when parental care strategies are not carefully considered for each species.Although maternal care is universal among mammals, the occurrence and extent of paternal and alloparental care varies substantially among species.

Many mammalian females require complete isolation from conspecifics and seclusion from potentially stressful events around the time of birth and early rearing. This appears to he particularly important for species in which only maternal care is provided. Maternal neglect, abandonment, or cannibalism of young has historically been one of the most common causes of death among captive-born mammals. Indeed, it seems that a large portion of unusually high infant mortality rates reported for several captive-bred mammal species may be attributed to parental care problems.

In tree shrews, for example, it has been documented that about 50 per cent of young are lost due to maternal neglect or cannibalism. Changes in environmental factors such as the provision for more than one den site can provide solutions to maternal care problems in some species. For example, in the wild, cheetahs and several other carnivore species move their young when disturbed.

Disturbance in captivity, such as a thunderstorm or other unpredictable events, may cause a similar reaction in the female. If only one den site is provided, cubs may be carried around excessively and eventually die of hypothermia or bite wounds. Behavioural studies on tree shrews showed that females will not share a nest with their young, but rather visit them and then return to their own den. Provision of additional nest-boxes in captivity therefore led to an increase in breeding success. In several ungulates where mothers spend a large amount of time away from their young in the wild appropriate hiding places are necessary to facilitate this separation of mother and infant. Reportedly, injuries and maternal care problems may result if constant visual contact of mother and infant is enforced due to inappropriate enclosure design. Maternal incompetence has been reported for a number of species and may result from lack of experience. Recommendations for captive management have been provided based on these findings.

For example, in bottlenose dolphins *Tursiops truncatus,* it has been suggested that first-time mothers should be kept with experienced females for enhanced rearing success and in gorillas and chimpanzees considerable effort has been undertaken to teach mothering skills to females. Overall, parental care behaviour appears to be a particularly labile function and can be disrupted easily even at low levels of stress when other aspects of reproductive function are normal. Hand rearing of captive-born offspring has often been employed to bypass some parental care problems encountered in many species.

For example, Nadler reported that more than 80 per cent of captive-born gorillas were hand reared. Similarly, in maned wolves *Chrysocyon brachyurus,* the majority of offspring have been hand reared because captive females rarely care for their young. However, this form of intervention may further complicate the longer-term prospects for obtaining a self-sustaining captive population. In some species, where paternal care or alloparental care are part of the natural care pattern, the presence of conspecifics may be necessary for successful rearing.

For example, the presence of the male seems to be essential for successful captive rearing in bush dogs. Survival of young was also highest for green acouchi *Myoprocta pratti* when male and female were present. Dudley showed that pups of mice

Peromyscus califomicus parasiticus were heavier during early development when the male was present. In a number of mammal species females rear their young cooperatively or communally. Colobine monkeys, for example, often transfer their young soon after birth to other conspecific females for prolonged periods of time. Alloparental care is also common in canids, rodents, bats, dolphins, and many other species. Such helping experiences may be important for the adequate development of mothering skills and may be crucial for reproductive success.

EARLY DEVELOPMENT AND SOCIOSEXUAL COMPETENCE

Transmission of intraspecific, learned behaviours from one individual to another is an essential part of early development in mammals. Experience with captive breeding of many mammals suggests that behavioural anomalies can result from rearing individuals in socially deficient environments. Social deprivation during early development is an underlying cause for breeding and parental care problems in captive-bred species.

For example, social deprivation of infant gorillas has led to deficits in social, sexual, and parental behaviour, and these effects appear to be difficult, if not impossible, to reverse. Social access during the first year of life is associated with higher reproductive success, and mother-reared females may be reproductively more successful than handreared ones. Hand rearing has been documented to result in behavioural deficiencies in adult individuals of several mammals. This practice may lead to a vicious cycle by producing more behaviourally deficient adults, which may either not breed or not care for young and therefore necessitate further hand rearing of offspring.

Furthermore, even in cases where maternal care is provided, premature removal of young from their mother, peers, or family group may have a negative impact on sexual and social competence because development may not be completed at the time of removal. Deficiencies in motor patterns may also result from inadequate social exposure during developmental stages. Improper orientation and mounting by males and failure to assume the lordosis posture in females are some of the problems observed in a variety of mammals. In addition to the social environment, the complexity of the physical environment can have far-reaching effects on the developing animal. Both neural and muscle tissue have certain "windows of vulnerability" during which they are particularly sensitive to environmental disruption. Appropriate stimuli are necessary to facilitate optimal development. Laboratory research, mainly on rodents, found a number of morphological, physiological, and behavioural consequences as a result of rearing in more complex environments. Some of these consequences include changes in brain morphology and differences in size, number, and complexity of neural synapses. Important behavioural changes include greater behavioural diversity and complexity when exploring objects.

Such differences in exploratory behaviour appear to reflect functional differences in learning and may be important for an animal's ability to behave adaptively when faced with a novel situation. Lack of environmental complexity may lead to severe behavioural deficiencies and development of rigid and abnormal behaviour patterns (stereotypic behaviours). For example, when reared in enriched environments, black-footed ferrets were able to kill hamsters more efficiently than individuals reared in less complex environments.

As a consequence, it has been suggested that animals reared in deficient environments may also be less able to cope with novel and stressful situations such as mate introduction and caring for young.

STRENGTHS AND WEAKNESSES OF BEHAVIOURAL RESEARCH FOR SOLVING CAPTIVE BREEDING PROBLEMS

STRENTHS

Behavioural research can be noninvasive, of relatively low cost, and can be carried out by trained zoo personnel, making it an ideal first approach to management problems. The provision of an environment in which natural (or close to natural) behaviours can be exhibited enhances the educational and research potential of a captive population. In addition, this practice is likely to decelerate adaptation to the captive environment and to increase breeding success, which in turn will permit the establishment of self-sustaining captive populations. Insufficient knowledge about husbandry and behavioural requirements are frequently the main impediments to successful captive propagation for many mammals. Systematic behavioural studies on wild and captive specimens can therefore allow us to provide the appropriate environment for a species to breed naturally (unassisted) in captivity. This will also facilitate the development of behavioural competence and preservation of the behavioural repertoire associated with breeding and courtship.

Learned and culturally transmitted behaviours are frequently an important part of early development in most mammal species. Although some behaviours may require little experience to be exhibited, others need to be acquired during various stages of early development in more or less complex social groupings in interactions with parents, siblings, and other conspecifics. Such behaviours may be lost permanently if no provisions are made for their preservation. Behavioural research can help identify problems and allow us to provide the appropriate social and physical environment for successful development. Finally, behavioural studies are necessary for reintroduction efforts. Appropriate social structure for release, foraging skills, predator avoidance, orientation, and dispersal are some of the important behavioural strategies that need to be examined carefully before any reintroduction attempts to enable the provision of adequate prerelease training.

WEAKNESSES

Behaviours may vary widely over time and between individuals. A trade-off between time spent investigating individuals in one area and overall number of individuals is necessary. Therefore, results are frequently based on relatively small sample sizes and on data obtained from one facility. However, innovative research approaches employing a variety of techniques, such as keeper questionnaires, remote sensing, and video recording, in combination with direct behavioural observations, can provide data from multiple institutions over the same time period.

Nevertheless, great care is required to standardize behavioural studies across individual researchers and facilities. The high variability in behaviour makes it difficult for behavioural studies to generate simple guidelines comparable to demographic and genetic guidelines for captive breeding. Instead, a species-specific approach will have to be applied, and even within a species, individual variation and individuals' needs may have to be considered to achieve reliable reproduction in captivity.

For example, individual cheetahs in captivity showed quantifiable behavioural variation with regard to fearfulness and other characteristics. A correlation was found between the degree of fearfulness and breeding status. Nonbreeders scored significantly higher than breeders on the fear component. Studies on domestic dogs and laboratory animals have shown a high heritability of fearfulness and selection may therefore occur for such traits in captivity. Temperamental differences and patterns in individual behavioural variation may therefore require consideration to counteract the adaptation to a specific captive environment. The captive environment may need to be adjusted accordingly to allow different types of individuals to reproduce successfully in captivity.

Behavioural studies are time intensive and often do not produce conclusive short-term results. Alone they may not provide all the information necessary to understand behaviourally based captive management problems such as mate choice, reproductive suppression, stressrelated problems; an interdisciplinary approach may be advisable in most cases. This chapter generated six specific recommendations.

First, appropriate enclosure design and furnishing are often established based on anecdotal and descriptive information and frequently on scant field data. More systematic investigation of the effects of various enclosure features on reproductive behaviour and parental care are urgently needed for numerous captive-held species.

Second, the dynamics of interspecific competition and the trade-off of close coexistence among different species in the wild and their possible ramifications for captive management have rarely been systematically examined and are in need of further investigation. In addition, the effect of housing prey and predators in close proximity needs further attention. Considering the preservation of naturally occurring behaviour patterns, it may be beneficial for prey species to have some degree of exposure to predators as long as ample hiding spaces and room for retreat are provided.

Third, studies on wild and captive populations of many species are necessary to identify species-specific foraging patterns and to investigate their relationship to other important behavioural variables, such as reproductive behaviour and parental care.

Fourth, there are many captive species in which problems associated with mating, incompatibility, and unknown reproductive behaviour are regarded as a major impediment to captive propagation. Although behavioural studies have provided solutions for some of these breeding problems, an understanding of underlying mechanisms and causes has not been achieved for many captive species, and behavioural-physiological research is urgently needed in this area.

Fifth, establishing time budgets and behavioural repertoires for wild populations is needed to provide a baseline for comparison to behaviours and time budgets found in captivity and to allow identification of abnormal behaviour patterns.

Sixth, combined genetic and behavioural studies identifying parentage in wild populations can help to establish the true mating systems of a species, which may in some cases differ from previous findings based on behavioural observations alone.

The above recommendations are based on the literature and are intended to provide a few salient examples of necessary future research. Many more could certainly be added to this list. In addition, four general recommendations can be made.

First, information from descriptive and anecdotal accounts on "do's and don't's" in captive breeding needs to be synthesized for each species and used to generate testable hypotheses.

Second, the development of standardized methods and training procedures for behavioural research in captive studies is necessary to allow for broader application of results. These methods should then be applied across facilities and countries. In particular, multi-institutional studies and innovative methodological approaches to the study of behaviour should be encouraged.

Third, integration of behavioural research with other research disciplines is becoming increasingly necessary. Interdisciplinary studies, for example, combining behaviour and physiology (increasingly feasible through the recently developed noninvasive techniques of fecal, urinary, and saliva steroid monitoring), will allow us to test hypotheses about complex aspects of behaviour and the interplay of behaviour, physiology, and various environmental factors.

Finally, guidelines and provisions concerning behavioural management of endangered species need to be an official part of captive-breeding management plans. Consultation with behavioural ecologists should occur at the beginning of any planned breeding Programme.

6

Behavioural Issues Involved in Birds

The success of conservation measures taken in the wild has been arbitrarily defined by Beck *et al.* as the establishment of a population in the wild of "at least 500 individuals which are free of provisioning or other human support," or where a "formal demographic analysis," *e.g.*, a population viability analysis, "predicts that the population will be self-sustaining." Here the goal is to increase population *numbers* rather than to improve the quality of individuals. A number of behavioural measures have been used to bolster populations in the wild.

PROTECTION FROM PREDATION

When the population size of a prey species has declined to dangerously low levels, predation may wipe it out. This has occurred frequently when native species are faced with exotic predators with which they have no previous experience or with which they have no history of potential coevolution. Many of the most famous examples are flightless birds on oceanic islands.

Hence, shielding vulnerable individuals of an endangered species from predation has often proved a straightforward means of protection. Such protective methods have taken the form of guarding birds' nests, providing predator-safe nest-boxes, or even placing nest-boxes in ways that exploit the territoriality of other species. The pearly-eyed thrasher *Margarops fuscatus* attempts to use the nest holes of the endangered Puerto Rican parrot *Amazona vittata*. In so doing, the eggs or chicks of the parrot are killed and eaten. By providing an extra nest-box near a parrot's box, the thrasher can be induced to occupy it without harming the parrot's brood. Moreover, through its territorial behaviour the thrasher keeps other thrashers away from both nests. Thus a detailed knowledge of the thrasher's behaviour helps protect the target species' brood. Knowledge of the behaviour of other nest competitors has aided in designing still other countermeasures, thus boosting successful reproduction of the parrot. Island faunas are among the most vulnerable, as evidenced by extinction rates and the main reason for extinctions has been the presence of introduced animals. Various transfer methods have proved useful in avoiding introduced predators.

According to a recent catalogue these methods are: reintroductions or attempts to establish a species in an area which was once part of its historical range from which it has been extirpated or become extinct; translocations or deliberate movements of wild individuals or populations from one part of their range to another; reinforcements or supplementations (the addition of individuals to an existing population of conspecifics); and "benign" introductions (attempts to establish a species outside its recorded distribution but within an appropriate habitat and ecogeographical area).

PROTECTION FROM INTRASPECIFIC BROOD PARASITISM

In many insect and bird species, females lay their eggs in other females' nests. Extra parental effort jeopardizes the reproductive success of the parasitized broods. As Semel *et al.* and Eadie *et al.* have shown for wood ducks Aix sponsa, judiciously placing nest-boxes in more secluded places and spacing boxes farther apart can markedly reduce intraspecific brood parasitism.

PROTECTION FROM SIBLICIDE AND RESCUE EFFORTS THROUGH CROSS-FOSTERING

Right after hatching, many species of birds of prey and herons kill their sibling(s), and their parents do not intervene.

In pioneering experiments, Meyburg cross-fostered endangered birds of prey and thus increased their fledging success. Due to the presence of conspecific nestlings before and after cross-fostering, misimprinting may be of minor concern. Since its discovery by Baron Adam von Pernau in 1702, imprinting has been widely studied, but its relevance for captive breeding has been appreciated only recently. For instance, it becomes a problem when threatened species are cross-fostered.

Whooping cranes *Grus americana* that had been raised by sandhill cranes *Grus canadensis* migrated and even flocked with conspecifics, but they failed to breed properly. When Chatham Islands robin *Petroica traversi* numbers had been reduced to one female and four males, cross-fostering the offspring to the Chatham Islands tit *P. macrocephala* was attempted. Misimprinting on the wrong species, a real threat to pure breeding of the robin, was countered by taking the focal species and its foster species out of reach of each other.

FORESTALLING INFANTICIDE

Both in the wild and in captivity, infanticide can decrease numbers of offspring reaching independence. Infanticide has been recorded in a wide variety of species and occurs as part of a species' normal behaviour that is easily absorbed in a large population. Because conservation efforts in the wild aim at increasing numbers, it may be unwise to remove potentially infanticidal individuals. As yet we have not developed methods of helping individuals defend themselves against infanticidal attacks.

CONSERVATION OF TOP PREDATORS

Top predators tend to depress the densities of their prey and of predator species at lower trophic levels ("mesopredators"); mesopredators in turn depress prey at lower trophic levels. The top predator therefore assumes the role of a "keystone" species in the community and its removal leads to losses of still further species. For example, the removal of top carnivores from a community benefits the mesopredators to the point that their prey come under strong predation pressure. Eventually the latter may be driven to extinction. Removal of coyotes *Canis latrans* boosted the populations of both gray foxes *Urocyon cinereoargenteus* and feral domestic cats so that the songbird prey of these latter mesopredators suffered accordingly. Consequently, it has been argued that keystone predators should receive more conservation attention than other members of the community. The same reasoning applies to "keystone mutualists" such as pollinators or seed dispersers.

BEHAVIOURAL ISSUES IN CAPTIVITY

For introductions of any sort to be successful, there must be large numbers of animals to be introduced, preferably of high *quality*. Assessing an individual's phenotypic and genotypic attributes is difficult, however, and in reality managers will have to be satisfied with an assessment of the health and nutritional status of captive animals that will be released.

ENVIRONMENTAL STRESS IN CAPTIVITY

Generally, removal of stress can promote breeding in captivity. For example, Shepherdson has emphasized the pivotal role of environmental enrichment in captive breeding. It may well be that this factor exerts its beneficial effect via a general reduction of fear of predation because enrichment entails providing objects that subjects can use for hiding and feeling safe. Zoo animals that are continually disturbed by visitors or that have a constant view of predators may not breed.

Thus caretakers should behave unobtrusively or even engage in friendly relations with their subjects. Similarly, the removal of threat from natural predators may have the same effect. For example, a pair of pigmy owls *Glaucidium passerinum* that failed to breed for several years bred immediately once it was shielded from view of a pair of eagle owls *Bubo bubo* in an adjacent aviary. However, the generality of this observation is open to question. A pair of great tits *Parus major* bred with impunity in an unoccupied nest-box of a captive pair of Tengmalm's owls *Aegolius funereus,* flying freely in and out the aviary. Habituation to a predator may at times overcome fear that would typically suppress reproduction. The evidence for or against habituation to captive predators in captive prey subjects is ambiguous.

Under natural conditions, fear of predators does not abate as a consequence of frequent stimulation, whereas fear of novel objects or places readily habituates. By contrast, under captive conditions some animals seem to lose their fear of predators, though this has not been explored systematically. Some animals may assess that predators constrained in enclosures or aviaries pose no real threat. Klinghammer observed similarly that hand-raised birds and mammals "are less fearful of many other stimuli and situations than normally raised animals."

He cautions that this deficiency renders hand-raised animals more vulnerable to accidents in new surroundings. In an insightful but widely ignored paper, Barlow drew attention to the fact that to behave normally (*i.e.,* similar to that in nature), animals need a moderate level of stimulation with both complete isolation and full exposure leading to fearful behaviour.

He eliminated stress in captive pairs of fish by presenting to them "dither" fish (or birds) behind a transparent partition. Treated this way, reluctant pairs started to breed successfully. Finding a balance between moderate levels of stimulation and reducing fear is a challenge for zoo breeding Programmes. Other beneficial effects of environmental enrichment, apart from breeding, include successful killing of prey after release.

FREE CHOICE OF MATES

In captivity, potential mates are chosen on the basis of genetic and demographic considerations for both maintaining maximum genetic diversity and avoiding inbreeding. These needs may directly conflict with free choice of mates. Free choice is often the key to success in propagating a target species, or enhancing its breeding success (*e.g., Falco sparverius*).

For example, in the monogamous cockatiel *Nymphicus hollandicus*, freely mated pairs had higher breeding success than force-paired birds. Nevertheless, prior association of force-paired mates before the onset of breeding conditions (long days) improved breeding success; pairs were better synchronized and performed breeding activities in less time. Species may be differently susceptible to forced pairing, ranging from the detrimental effects mentioned to immunity in other species mammals seem less affected than birds. Long-lived birds that form life-long pairs are expected to be more choosy than shortlived species in selecting a mate.

This is because errors committed in selecting the "wrong" mate multiply with each bout of breeding. "Divorce" may be preceded by lower breeding success), suggesting that errors committed in pair formation can be corrected. This possibility is largely denied to the majority of birds that are short lived. Ideally, potential pair mates should gauge their choice of partner on both compatibility and mate quality. There is scant evidence for either. Failed breeders tended to fail because they did not coordinate their activities well in time and female blue tits *Parus caeruleus* seem to gear their rate of extrapair copulations to the quality of their mates as measured by their mates' survival.

But how do potential mates assess compatibility and/or quality in advance? In a valiant effort to discover which of 13 morphological features females use in mate choice, Ostermeier found in the sexually monomorphic Javanese mannikin *Lonchura leucogastroides* that any one or a combination of three morphological characters and older age rendered a male successful against a rival in choice tests. The results might be interpreted in many different ways. Some of the morphological traits, including body weight, correlated significantly with male age.

It is plausible to assume that what females were after in a mate was older age and perhaps heavier weight, and that the characters chosen were merely indicators of age. Age is assumed either to be an indicator of "good genes" or of being a good father because of greater experience. This may be true up to a point but the choice of an older male may be compromised by its senescence impairing its parental competence. Domestic pigeons *Columba livia* do in fact exhibit morphological signs of senescence at old ages, and old males seem to be less often chosen by females than males of prime age, although they should excel in breeding experience. In the mannikin study female choice was based on multiple criteria as judged from the female population as a whole. Similar results have been obtained in the sexually dimorphic guppy *Poccilia reticulata.*

In both cases some condition-dependent criteria (weight and age in the mannikin; courtship and dominance rank in the guppy) may allow females to gauge male quality. The "redundant signal" hypothesis suggests that several traits in combination provide a better estimate of general condition than does any single ornament. Although conceived for explaining multiple ornaments, the hypothesis can also explain rules of pair formation in a monomorphic (*i.e.*, ornamentless) species, as shown by the mannikin study. In laboratory mice inbreeding is avoided by olfactory cues connected to the major histocompatibility locus. The odours linked with this highly variable locus offer to a choosy mate a convenient handle by which it can compare its own odour ("self") and, hence, genes, with the odour of a potential mate. As a result, phenotype matching translates into genotype matching with the aim of choosing, in this case, a genotype least susceptible to inbreeding. So far, not a single study permits one to predict which individuals of a species would or would not make a good pair. This is because multiple criteria may underlie female choice, and still others have surely been overlooked by researchers. In species with pair bonding, the prediction is even more difficult: even if the criteria of choice were known, the necessary harmony among pair mates is an emergent property of the phenotypic profiles of both.

REMOVING REPRODUCTIVE SUPPRESSION

At times, the reproductive potential of a group of potential breeders cannot be realised because of reproductive suppression by high-ranking group members. This occurs, for example, in golden lion tamarins *Leontopithecus rosalia*. Artificial fissioning of groups may help to alleviate this impasse, thus permitting suppressed yet physically mature animals to reproduce.

PREVENTING SIBLICIDE

Captive breeding of birds in zoos can be severely marred by siblicide. In the endangered bearded vulture *Gypäetus barbatus*, the older of the two chicks habitually kills the second hatching sibling. Because siblicidal aggression wanes within a few weeks, fledging success could be markedly enhanced by placing a baffle between the two chicks.

HAND-REARING AND SOLUTIONS TO ITS DRAWBACKS

Captivity often impairs the quality of animals and adversely affects their potential for release. In particular, artificial rearing by hand or by a foster-parent or surrogate, a popular means of increasing population numbers, exacerbates this problem. It produces animals that are often incompetent breeders and hence perpetuates the very problem that gave rise to it. On the other hand, hand-rearing may be essential for keeping a species in captivity at all. For example, certain birds can only be familiarized with their artificial diet by being raised on it and green iguanas *Iguana iguana* acclimate to captive conditions well only when growing up under them.

REPRODUCTIVE ABNORMALITIES THROUGH HAND-REARING

In a systematic study, Myers *et al*. found that hand-reared cockatiels had a strikingly lower breeding success than did parent-raised birds. Although hand-reared females laid more eggs, they failed to fledge more young because they misplaced eggs on the cage floor. For breeding to be successful, a parent-raised male had to be present, but this condition was not always sufficient. Absence of the right nest or appropriate building materials during hand-rearing may later inhibit constructing nests as adults. In some pairs the behaviour deficit can be remedied by one failed breeding experience. For reasons unknown, some females that had laid first on cage floors used nest-boxes in their second cycles, and hand-reared males that had been infertile produced fertile eggs. In addition, the type of nest-box used affected reproductive output of cockatiels in several ways.

Breeding in cavity-type boxes (with a normal entrance hole) as compared to shelf-type boxes (with an open front) improved fertility, hatchability, and fledging success. Despite this inferiority of shelf-type boxes, they may facilitate subsequent breeding. These results warrant further experiments and inclusion of other species, for they have implications for the propagation of cavity-nesting species and other birds. In mammals, the ill-effects of hand-rearing pertain to impaired behaviour during birth and mothering. Hand-reared males of certain primates orient themselves in the wrong way and show other deficits during copulatory behaviour.

IMPAIRMENT OF PREDATOR AVOIDANCE AND ITS RELATION TO MISIMPRINTING

Predator avoidance mechanisms in animals are either innate (*i.e.*, require no individual experience with the predator in question) or require learning of various sorts, or a combination of the two. Artificial rearing of birds and mammals, in particular, tends to disrupt the normal development of innate enemy recognition. There is no problem as long as captive stock is held in captivity, but predator recognition and avoidance becomes an issue when captive stock is released into the wild.

Klinghammer showed that ring (barbary) doves *Streptopelia roseogrisea* develop fear of humans when raised by their parents, but are tame ("blind" to a human predator) when raised by hand before they are 8 days old. A second effect of this early hand-rearing is misimprinting on humans as the objects of later sexual responses. The two effects may be the outcome of the same underlying mechanism, as the fear of humans is shown only after the young are mature at 8-9 months of age. In mourning doves *Zenaida macroura*, misimprinting is similarly affected by postfledging experience. When removed from their parents by their 8th day of life, kept in isolation until 52 days of age, and then given access to conspecifics in a community cage, mourning doves invariably mated with their species' members and rejected humans in a free-choice test.

Furthermore, mourning doves that are misimprinted on humans can be made to shift their sexual responses to conspecifics if given continual access to doves across four breeding seasons. Hence, experience with species' members after 52 days of age can still reverse the adverse imprinting on humans and forestall any fear of species' members. Yet the abolishment of fear of humans caused by hand-rearing is permanent in both species of doves. The "evil duo" of predator blindness and misimprinting is, at first sight, bad news for planners of release Programmes. However, it may be tempered first by the observation that ring doves lose their tameness towards humans immediately in strange surroundings, as do ravens *Corvus corax*.

Second, when hand-reared with a sibling, ring doves do not become misimprinted on humans. Although this protocol does not work with ravens and jackdaws *Corvus monedula*, it does work with parrots, where sibs can even be replaced by other species' nestlings (*Amazona* spp. with *Psittacus erithacus* and vice versa) (W. Wüst, personal communication).

Third, in other species of birds, imprintability may be more malleable than in the two dove species. In ravens, imprintability depends on the length of exposure to the imprinting object. Also, a classic criterion of imprinting, irreversibility, has recently been found to be less stringent than previously thought.

Fourth, it has been suggested (E. Thaler, personal communication) that in hand-reared grouse (*Alectoris graeca, Lagopus mutus, Bonasia bonasus*), the innate fear of humans will be maintained into adulthood when only a single human caretaker has raised them and towards whom the birds have no fear. This can be altered with more caretakers. A similar situation has been observed at crocodile farms although these animals perceive humans in the wild rather as prey than as a predator. A further caveat concerns the number of caretakers for foster-parenting animals. In species with biparental care, imprinting may become established more firmly if two caretakers instead of only one are involved. The lesson to be learned from these studies is twofold.

First, there may be a critical point in time when exposure to a species renders the animal perceptually blind towards potential predators; that juncture may or may not be related to sexual imprinting depending on the species involved.

Second, during development of their animals, captive-breeding specialists should minimize contact with species that may act both as imprinting objects and potential predators. Remedies include hand-rearing individuals in groups or with other species or with a mechanical model parent and in the group-raised category with conspecifics or with other species. These refinements need to be tested with species from various taxonomic groups.

BEHAVIOURAL ABNORMALITIES DUE TO FOOD DEFICIENCIES

Although nutrition has been at the center stage of animal husbandry for a long time, its role is just now becoming more appreciated by conservationists. There is anecdotal evidence that a deficient diet for young birds may result in death when going through the first molt of the flight feathers or result in the loss of the whole brood due to impaired parenting. This is despite the fact that full-grown, ill-nourished birds may actually look healthy to an expert's eyes.

The first systematic study exploring the ill effects of a deficient diet on behaviour was conducted with European blackbirds *Turdus merula* by Pollack. She raised two groups of nestlings from age 4 days to independence, with one group receiving an optimal diet and the other a diet deficient in terms of six species of insects and rennet. The two diets were otherwise identical with regard to animal protein, carbohydrates, and snail shells from day 7 on, and body weights did not differ significantly.

After fledging the restricted-diet birds displayed reduced neophobia, greater tameness towards their human foster-parent, increased intraspecific aggression, and lower activity in terms of location calls, vigilance, and flight behaviour. Similar though less varied behavioural changes have been found in a similar food deprivation experiment with great tits. Two interrelated questions remain: do the behaviour changes persist into adulthood, and can they be remedied later on by supplementary feeding?

The lesson from these still isolated findings for the captive breeder is clear: aside from the many direct effects of hand-rearing on behaviour, indirect effects operating through nutrition must now be heeded. Apart from being detrimental to any release Programme, a deficient diet may also hold the key for improving captive breeding.

REINTRODUCTION OF CAPTIVE-BRED ANIMALS

Releasing animals from captivity into the wild can be classified as "hard" when introduced animals have no prior familiarization with the release area, and "soft" when the animals have some familiarity with the area before final release. There are many problems faced by managers in both sorts of reintroductions.

EXPLORING THE ENVIRONS NEEDS TUTELAGE

One reason for the poor success of releases is that introducing large numbers of animals precludes allowing them to explore the release area under the tutelage of their parents. In a pioneering study, Thaler *et al.* developed a successful protocol with hand-reared bald ibis *Geronticus eremita* that took the long-lasting family ties of this species into account. Continuous guidance of only six hand-reared fledglings by two alternating human caretakers familiarized the birds with an ever larger area around their acclimation cage in the countryside. This tutelage continued until the birds were entirely independent. In this way the panic flights seen in hundreds of captive-bred bald ibis precipitated by hard releases in the Near East were avoided. Behavioural deficiencies that hindered the release of golden lion tamarins were alleviated by providing them with wild-born individuals as tutors.

EXPERIENCE WITH LIVE PREY AND OTHER FACTORS FACILITATE HUNTING

Raptors and carnivores often need to be trained to hunt. In captivity, opportunities for such training are restricted so that captive-born predators often lack the necessary skills to recognize, locate, subdue, and handle their prey. Wild avian and mammalian predators increase their hunting success dramatically with age. In prerelease training, individuals need to be primed (pre-exposed to prey) to perfect hunting skills and reinforcement must occur through killing.

Dietary cues can also be transmitted via olfaction or even through conspecific alarm calls. When captive-born eagle owls are given live avian and mammalian prey before 220 days of age, they hunt their full array of prey when later released. If denied live prey until that age, they refuse to kill these prey later on.

Releases can therefore take place only when live prey has been introduced early enough (*i.e.,* between day 110 and 220). Also, release of eagle owls needs to occur before a critical period of live prey-hunting ends.

In mammals, many factors affect the development of killing behaviour. For example, in polecats *Mustela putorius,* social play with litter mates, maturational processes, and a critical period to exercise the neck bite may all interact in the perfection of the killing response. Imprinting on olfactory and other stimuli associated with the prey may further complicate the matter. In the endangered black-footed ferret *Mustela nigripes* with no killing experience before release, environmental enrichment of maintenance pens contributed to the proper killing of hamsters after release. At present there are no general rules for turning captive-born predators into competent hunters.

PREDATOR AVOIDANCE MAY REQUIRE LEARNING

Many released animals suffer unnaturally high mortality from predation. It is therefore imperative that one understands why captivity renders introduced animals so vulnerable to predators even after being raised by their parents. There are at least three ways in which the social environment affects proper antipredator behaviour, while the predator itself may instill avoidance as well.

PSEUDOCONDITIONING

Juvenile three-spined sticklebacks *Gasterosteus aculeatus* fail to avoid pike *Esox lucius,* an important predator, when raised in captivity without their fathers. When sticklebacks from pike-threatened populations are reared by their fathers, they develop pike avoidance even in the absence of experience of pike. The mechanism underlying pike recognition is unknown, but it may be that when fry are retrieved from the nest by their father, they experience an aversive stimulus that they later generalize to the pike. Because being reared without a father does not impair the escape response per se but rather the perceptual system underlying it, the normal development seems to imply pseudoconditioning rather than sensitization.

IMPROVEMENT AND MAINTENANCE OF ALARM RESPONSES THROUGH PARENTS

Many birds and mammals deploy innate antipredator behaviour. Yet the social environment modulates such responsiveness in ways crucial for release projects. The effects of hand-rearing versus parent raising range from a complete failure to develop appropriate alarm responses, to maturing later, to needing longer stimulation or disappearing rapidly.

Decay of responsiveness to a predator demonstrates that the presence of or guidance by the mother is indispensable for the maintenance of the response, as, for example, in hazel grouse *Bonasia bonasus.*

In no instance has hand-rearing led to a response that is superior to that of hen-raising. Yet the ways in which the presence of the parent hen brings about improvements have gone unanalysed so far. The effects of the parent hen must be rather specific. Bantam hens, a domestic breed popular as foster parents among pheasant breeders, fail to instill in gray partridge the appropriate antipredator and avoidance behaviour.

Equally alarming is the fact that there are differences among the maintenance procedures of different captive breeders. Extraordinarily, in black grouse, the behaviour of handreared chicks towards a fox was more similar to chicks raised by a hen-reared hen than to chicks raised by a hand-reared hen. The hand-reared hen was more docile and more chick-directed, whereas the hen-reared hen was more fearful and rarely gave contact or warning calls.

Therefore, the chicks of this latter hen grew up more similar to the hand-reared chicks. This indicates that hand-rearing may produce a good mother but behaviourally handicapped chicks.

Hence, the history of the parent may be of paramount importance to any prerelease training. Ignorance of the precise mechanisms underlying the normal development of antipredator behaviour has precluded any successful mass-breeding technique and has thus prevented mass release Programmes from being successful. The outcome of normal development should be calibrated against the behaviour of wild-caught, well-acclimated adults.

CULTURAL TRANSMISSION VIA CONDITIONING TO PREDATOR STIMULI

When innate enemy recognition does not occur, animals can learn in other ways which lend themselves well to prerelease training. Juvenile three-spined sticklebacks *Gasterosteus aculeatus* fail to avoid pike *Esox lucius*, an important predator, when raised in captivity without their fathers. When sticklebacks from pike-threatened populations are reared by their fathers, they develop pike avoidance even in the absence of experience of pike. The mechanism underlying pike recognition is unknown, but it may be that when fry are retrieved from the nest by their father, they experience an aversive stimulus that they later generalize to the pike. Because being reared without a father does not impair the escape response per se but rather the perceptual system underlying it, the normal development seems to imply pseudoconditioning rather than sensitization.

IMPROVEMENT AND MAINTENANCE OF ALARM RESPONSES THROUGH PARENTS

Many birds and mammals deploy innate antipredator behaviour. Yet the social environment modulates such responsiveness in ways crucial for release projects. The effects of hand-rearing versus parent raising range from a complete failure to develop appropriate alarm responses, to maturing later, to needing longer stimulation or disappearing rapidly.

Decay of responsiveness to a predator demonstrates that the presence of or guidance by the mother is indispensable for the maintenance of the response, as, for example, in hazel grouse *Bonasia bonasus*. In no instance has hand-rearing led to a response that is superior to that of hen-raising. Yet the ways in which the presence of the parent hen brings about improvements have gone unanalysed so far.

The effects of the parent hen must be rather specific. Bantam hens, a domestic breed popular as foster parents among pheasant breeders, fail to instill in gray partridge the appropriate antipredator and avoidance behaviour. Equally alarming is the fact that there are differences among the maintenance procedures of different captive breeders. Extraordinarily, in black grouse, the behaviour of handreared chicks towards a fox was more similar to chicks raised by a hen-reared hen than to chicks raised by a hand-reared hen. The hand-reared hen was more docile and more chick-directed, whereas the hen-reared hen was more fearful and rarely gave contact or warning calls.

Therefore, the chicks of this latter hen grew up more similar to the hand-reared chicks. This indicates that hand-rearing may produce a good mother but behaviourally handicapped chicks. Hence, the history of the parent may be of paramount importance to any prerelease training. Ignorance of the precise mechanisms underlying the normal development of antipredator behaviour has precluded any successful mass-breeding technique and has thus prevented mass release Programmes from being successful. The outcome of normal development should be calibrated against the behaviour of wild-caught, well-acclimated adults. When innate enemy recognition does not occur, animals can learn in other ways which lend themselves well to prerelease training. Although repeatedly sugof others. Hence, there is "preparedness for learning" biologically relevant stimuli are more effective than others.

In blackbirds, the live conspecific can be replaced with taped alarm calls including those of other syntopic species. Training with the help of a surrogate therefore renders the experimental protocol of cultural transmission especially amenable to release training. An innovative approach that employed a related species as tutor was adopted by Carpenter *et al*. They released the endangered masked bobwhite *Colinus virginianus texanus* together with knowledgeable members of a nonthreatened subspecies, *C. v. ridgwayi*. As a result, masked bobwhites acquired the necessary skills in avoiding predators and finding food. The tutor species' members had been previously sterilized to forestall hybridization.

ACQUISITION OF PREDATOR AVOIDANCE VIA TRAUMATIC EXPERIENCE

A simpler, more straightforward way of acquiring knowledge of predators is to learn to avoid them as a result of being stared at, silently approached, or pursued at full speed. This relies on some perceptual mechanism decoding generalized threat stimuli that triggers avoidance behaviour. As a consequence of being targeted in this way, details of the entire stimulus become associated with its unconditioned aspects.

Reinforcement obviously arises from the performance of the avoidance response and as a consequence, predators may come to be avoided that originally were not recognized. In such an experiment, Curio produced fear of cats by having a live feral cat harass a caged Galapagos ground finch *Geospizafortis*.

Without this treatment this species and other geospizines are "blind" to cats; carnivores have never been part of the Galapagos fauna. In regions where prey have not coevolved with predators, such as New Zealand, predator blindness is common. Consequently, the native fauna suffers from introduced predators, and many species have become extinct. The takahe *Phorphyrio mantelli,* a flightless rail in New Zealand, has dramatically declined due to climate change and the introduction of stoats *Mustela enninea.*

The rail's evolution in a mammal-free environment has led to a loss of carnivore recognition. To prepare captive-bred takahe for release, Hölzer *et al.* trained 10 chicks to avoid stoats in their rearing pens. Birds were first tested with a stuffed stoat and then frightened using a "stick stoat," manipulated by a hidden observer, and a "string-stoat," suspended puppet-style from an overhead line, which was made to attack the chicks in various ways. This treatment was interspersed by two others in which the stoat attacked a replica of an adult takahe emitting distress calls until it was "killed." In another treatment designed to engender self-defence, the takahe replica strongly pecked at the stoat until it was ("killed.") As a result of these treatments, takahe chicks became more vigilant, approached the stoat significantly less, stayed farther away, hid more often, and showed more signs of discomfort, such as standing alert and making escape panic runs. Coping with a stoat by retaliation was not learned, however, although all birds had intensely watched the takahe replica hitting the stoat.

The length of time that these responses to the stoat would persist, to what extent they would persist after release into the wild, and how they would translate into enhanced survival have recently received a tentative answer. Captive-reared and released takahe that had been followed by radiotelemetry survived about as well as wild-reared ones.

However, the number of birds and years of study are small. Finally, the behavioural modification would have to be culturally transmitted across generations to have a lasting effect on the well-being of the released population. The same questions apply to the successfully conditioned avoidance of introduced predators (fox, cat) of the Australian rufous hare-wallaby *Lagorchestes hirsutus;* individuals were subjected to a similar treatment using dummies. Ellis and Serafin trained masked bobwhite quail for release by harassing captivebred individuals with dogs, humans, and a trained Harris's hawk *Parabuteo unicinctus.*

Before release, the birds showed greater general mobility, covey coordination, and predator avoidance skills. As a result of this prerelease training, released bobwhites survived better than nontrained birds. Only in this study has it been documented that survival is enhanced in the wild as a consequence of prerelease training of a bird. There may be a fundamental difference between training an animal that has lost (or never had) the ability to cope with a certain predator over its evolutionary history and an animal that has become predator-blind due to captive breeding. In captive animals, depending on the degree of their deprivation proper coping behaviour may be instilled through minute improvements of the rearing regime.

Yet this may not apply to an animal that evolved in a predator-free environment. For example, Darwin's finches on islands devoid of avian raptors show less fear of them, but they have a pronounced fear of snakes, and finch-eating snakes do occur on some islands. Perhaps faunas that have gone through extinction filters in the past have developed more resilience to novel threat for example, from introduced predators. As a case in point, more than 90 per cent of bird species that have gone extinct in historic times lived on oceanic islands.

These islands are characterized by a paucity of vertebrate predators. Because predation has played a prime role in the extinction process it appears that both innate coping mechanisms and avoidance learning via traumatic experience had been impaired or lost during evolution on the islands. In these species, learning by traumatic experience may be the only way to train individuals for release.

UNIDENTIFIED LEARNING MECHANISMS

Some very simple forms of treatment may prove useful

for releases. For example, laboratory mice of various strains that prove notoriously insensitive to predatory threat (*e.g.*, movement, noise etc.) turn almost wild by being allowed to dig burrows in natural soil.

After a few hours of digging and moving in and out of their burrows they respond with instantaneous flight below ground when a human approaches. Similar observations have been made with domestic guinea pigs *Cavia aperea* running free along tunnels in lawns. Further, lizards (*Agama* spp.) adapted to indoor maintenance conditions where they have been tame to their caretakers turn wild under outdoor conditions, receiving their caretaker with their mouth agape in defence. Behavioural changes accompanying both planned and accidental releases of animals are still unstudied.

STRENGTH AND WEAKNESSES OF THE BEHAVIOURAL APPROACH

Knowledge of a species' behaviour may be indispensable for management intervention to succeed. Such knowledge may save valuable resources and reduce mistakes in captive breeding and reintroductions. Understanding the function of behaviour is of great help in gauging the consequences of any prerelease training on released animals' survival and reproduction. It may be profitable to consult behavioural ecologists in this endeavor, as they are used to viewing behaviour as adapted to the species' environment and are therefore especially qualified to spot a mismatch between the ill effects of management and fitnessmaximizing behaviour.

Nevertheless, behaviour studies are time intensive, and rescuing a species is often a race against time. In both captive breeding and release Programmes, the cost of time may be minimized when intervention measures are tried out with nonthreatened species as proxies of the endangered species. Another difficulty is that behaviour studies conducted at different breeding facilities can be compared only with great caution because of lack of standardization across both observers and institutions although these problems can be ironed out in the long term.

RECOMMENDATIONS

While *in situ* activities are largely concerned with maximizing the numbers of individuals of a species, both *ex situ* and release Programmes need to attend to both numbers *and* the quality of breeding individuals or releases. Findings in this chapter suggest that Programmes should seek to encourage behavioural research both in the wild and in breeding facilities.

Zoo staff need to be convinced that allocating part of its maintenance facilities to behavioural research will benefit breeding or release Programmes. A number of specific recommendations follow from this. First, in regard to captive breeding of birds, stress from handling and exposure to predators should be removed and enclosures enriched. Noninvasive methods should be used in capture and handling. When breeding species in captivity, females should have access to a male of her choice.

This pertains only to situations where genetic screening has shown that a pair is suitable for breeding. Such a proviso needs to be relaxed when the captive population size is so low that propagation of whatever individuals must have precedence over not breeding the species at all. In addition, managers should remove sources of reproductive suppression and prevent siblicide. Reproductive suppression may be alleviated or avoided by finding the appropriate social structure conducive to breeding.

Finally, when subjects are hand-reared, steps must be taken to avoid behavioural abnormalities that may affect reproduction. In regard to captive breeding and release, hand-rearing should be avoided. Instead, subjects should be raised in the company of siblings or even other taxa if research suggests that misimprinting on those siblings does not occur. Before release, carnivorous species must be competent at killing live prey and releases should be afraid of potentially dangerous novelty and humans.

In particular, releasees must develop their innate fear of predators or be trained by some appropriate procedure with conspecifics or aversive stimuli. A sound knowledge of a species' behavioural ecology is important in three areas of conservation management: in the wild, in captive breeding (conservation breeding) and in prerelease training and release. Management of bird populations *in situ* may require protection from predation, prevention of intraspecific brood parasitism, siblicide and infanticide, cross-fostering, and sustaining predation at normal levels.

The ill-effects of cross-fostering may be overcome by withholding the species of the foster-parent to facilitate mating with conspecifics. Although it is a major objective of captive breeding, the establishment of endangered species in their native habitat has met with varying success. Major obstacles have been breeding sufficient numbers of releasees and release of individuals that are poorly adapted to the natural habitat, such as losing traditions of migratory routes or the readiness to migrate at all.

In captivity, removal of stress facilitates breeding; this includes shielding prey from seeing predators. Free choice of mates has been essential for successful breeding in some species. Hand-rearing or cross-fostering individuals often renders them unsuitable for release because of deficiencies in breeding, not recognizing or coping with food, or becoming misimprinted on the foster-parent and predator blind to humans. Behavioural deficits may be forestalled by raising the animal with siblings or chicks of related species, minimizing contacts with humans, or restricting contact to only one care-taker; or remedied by breeding with a normally raised mate, or by tiding over a period failed breeding.

The sexes may be affected by hand-rearing differently. Deficient nutrition may generate far-reaching but subtle ill-effects on behaviour. These deficits are among the least known and the most difficult to control for. In prerelease training and after release, experience with live prey during a sensitive period and environmental enrichment have been found to bring about appropriate hunting and killing behaviour in raptors and carnivorous species.

Similarly, lack of predator avoidance has perhaps been the most important behavioural deficit in released animals. An innate avoidance response (*i.e.*, one functioning without prior experience with the predator concerned) may need parental (maternal) guidance to be maintained or, perhaps, to come about at all. Both cultural transmission of enemy recognition and an individual's traumatic experience with predator models, and other learning paradigms, can produce predator avoidance.

To design an optimal pre-release training Programme, the adverse effects of captivity per se need to be separated from those of rearing animals in isolation. The strength of the behavioural approach lies in that it is cost effective, but its major weakness is its lack of generalizations across species.

MATING SYSTEMS AND CONSERVATION PROBLEMS

The way in which individuals acquire mates, the number of partners they acquire, their type of pair bonds, and the form and division of parental care are all components of a mating system. Differences in mating systems are therefore reflected in differences in many facets of animal behaviour, including male and female reproductive tactics aspects of parental care variance in reproductive success and adult sex ratios. Each of these factors has the potential to influence N_e, population growth rates, and hence the resilience of a population to exploitation and speed of population recovery.

The first two chapters in this section investigate some of these connections, while the third tests the importance of different aspects of mating systems in influencing N_e.The strength of these chapters, taken together, is in highlighting the diversity of ways in which animal mating systems are pertinent to problems in conservation biology and in suggesting numerous avenues of further research. At the most basic level, most animals have to come together to mate, and the mating system dictates the different types of association between individuals in terms of cooperation, permanency, and even location. Dobson and Poole address the ramifications of conspecific aggregations for several aspects of conservation biology first by drawing attention to the importance of conspecific aggregations for population demography. For years, ecologists have realised that population growth rates can be nonlinear with respect to population densities; Allee pointed out that there may be a critical density threshold below which a population may experience a precipitous decline. There are still few empirical examples of the Allee effect, however, and the mechanisms by which this could occur remain obscure. Next, Dobson and Poole show how the spread of disease is influenced by patterns of aggregation in a host population, with infection more likely to be transmitted when hosts live close together. Since diseases are an important but relatively unexplored topic in conservation and because populations will become increasingly tightly packed in isolated habitat patches, this topic needs further modeling and experimental work.

Dobson and Poole then go on to examine how selective harvesting affects a population's response to exploitation and its potential to recover. In their example from elephants *Loxondata africana,* ivory poaching removes first large-tusked males, then females, and finally juveniles from a population resulting in highly female-biased adult sex ratios. Because males gain access to mates by searching over large distances for clumped groups of related females and because females preferentially mate with older males, selective removal of largetusked males may prevent females from becoming pregnant. In addition, the distribution of female group sizes also affects the rate of offspring production, with fewer calves born as females aggregate in smaller and smaller herds. This story is now one of the classic examples linking animal behaviour and conservation. Finally, Dobson and Poole point to a growing literature on individuals using conspecifics as cues as to where to settle and breed. This has repercussions for both metapopulation dynamics, reintroductions, and the efficacy of reserves because all assume that a habitat is equally attractive whether or not conspecifics are present.

Vincent and Sadovy examine the ways in which seven different aspects of mating systems in marine fishes could affect responses to fisheries exploitation. They suggest that fecundity could be limited if larger fish were taken in nets, as size and fertility are often correlated; that fishing could disrupt spawning activity; that removal of fish with long-term pair bonds could disrupt offspring production; that the operational sex ratio might be biased if one sex is larger and preferentially taken; and that removal of males could impact paternal care of offspring.

In their last two examples of the ways in which individual adaptive strategies could affect populations in surprising ways, Vincent and Sadovy focus on species in which individuals show more than one reproductive tactic. In coho salmon *Oncorhynchus kisutch,* males either mature when small and young (jacks) or delay maturation (hooknoses) in a frequency-dependent manner. Because hooknoses spend more time at sea and are therefore at greater risk from fishing, jacks may be at a selective advantage under intense fishing practice.

Eventually this will alter the phenotypic distribution of the male population and will additionally lower fishing returns. In a second example of hermaphroditic fish species in which adult fish change sex either at a given size and age (uncompensated sex change) or in response to social conditions (compensated), Vincent and Sadovy compare how these different mechanisms will lead to different adult sex ratios in exploited populations. For example, in protogynous fish species, removal of large individuals (here males) may lead to femalebiased sex ratios in uncompensated systems but to equal sex ratios in compensated situations, which in turn could affect population growth rates in the remaining population.

In essence, Vincent and Sadovy suggest that models of population response to harvesting (or population recovery) will be too simplistic without considering the subtleties of mating systems.

7

Behaviour of Vertebrates

Vertebrates often are marked to facilitate identification of free-ranging individual animals or groups for studies of behaviour, population biology, and physiology. Marked animals provided data for many of the topics discussed in this volume, including home range use, resource selection, social behaviour, and population estimation. Markers can be classified into three general categories: mutilations, tags and bands, and radiotransmitters. The appropriate marking technique for a study depends on several considerations, including study objectives, target species, marker cost, marker efficacy, and marker effects on the animals.

Studies using marked animals are characterized by the assumption that marking does not affect animals or that negative effects are not important. The assumption of no significant marking effects is critical because it is the basis for generalizing data collected from marked individuals to unmarked animals and populations. However, the assumption has not been tested rigorously for most marker types or animal species, despite the often necessary use of seemingly invasive marking techniques. The general paucity of marker evaluation studies apparently is related to the difficulties associated with conducting such tests in the field, as well as the belief that marker evaluation is tangential to most study objectives and therefore of minor importance to the researcher.

In addition, studies that evaluate marker effects often suffer from small samples, thus leading to qualitative conclusions or weak statistical inference. Researchers tend to choose markers that intuitively seem least likely to induce abnormal behaviour or survival, even though data supporting that assertion usually are weak or lacking. However, if the assumption of no marking effects is violated and the effect is not evaluated, then data collected from marked animals will be biased. It follows that if significant marker effects remain undetected or unaddressed, conservation and management actions based on those results might not be appropriate. In addition, recent guidelines established by institutional animal care and use committees require that marking protocols minimize pain and stress to study animals.

If researchers collectively ignore the development, evaluation, and application of animal markers acceptable to such committees, and fail to publish results of studies not finding significant effects, then some research might be needlessly jeopardized or precluded. The effects markers can have on animals and to examine critically the treatment of potential marking effects by ecologists. We use the word effect to mean unusual or abnormal behaviour, an abnormal function, or abnormal reproduction or survival. We use significant to indicate statistical results and important to indicate an observed effect and implication for studies. We emphasize the shortcomings of various marking techniques to animal biology. Our discussion is restricted to effects of markers, and thus does not include a specific review of handling effects.

Furthermore, we do not present results specific to causes of pain or stress because essentially no data exist from wildlife. First, we present the variety of marking techniques that are available for, and explore possible implications of markers on, various taxonomic groups. Next, we review recently published articles to examine how researchers consider potential marking effects. Finally, we discuss how potential marking effects can be minimized and evaluated in future studies. Consistent with the theme of this volume, the approach we have taken is often critical of existing information and protocols. However, such an approach is necessary if researchers are to improve the overall quality of data being generated from ecological studies. Nietfeld described available marking techniques (excluding marking with radiotransmitters) and generally reviewed marking techniques for vertebrates (excluding fish).

Samuel and Fuller provided similar information about radiotransmitters. Stonehouse edited a book about animal marking, and other overviews dealing with selected vertebrate groups include Stasko and Pincock, Wydowsky and Emery, and Parker *et al.*for fish; Ferner for amphibians and reptiles; and Marion and Shamis, Calvo and Furness, and Bub and Oelke for birds. These sources will lead the reader to the literature dealing with many species, many marking methods, and various considerations associated with different techniques, different species, and study objectives.

It is worthwhile to reiterate some important factors that Nietfeld *et al.* and others noted as important when deciding which markers to use for a study. Expense can be an important consideration because marking materials can range widely in cost (*e.g.*, tags versus radiotelemetry via satellites). The procedures required to initially capture and mark animals and to obtain results from intensive field observations or recapture efforts also are important. Markers should be easily assembled and attached, recognized in the field, and durable enough to remain functional throughout the study.

Additionally, all marking techniques should result in minimum pain or stress to the animal during application and use. Finally, markers should not cause abnormal behaviour or affect survival. Clearly, it is difficult to address all these criteria satisfactorily before the initiation of a study, so some marking has undesirable effects on animals and research results. The adverse effects of marking often are speciesspecific and might occur only in conjunction with certain behaviour (*e.g.*, courtship) or environmental conditions (*e.g.*, extreme temperature). Also, the magnitude and importance of such effects are highly variable among marker types. We present examples of marking techniques and their effects on vertebrate species. This material will help address questions about adverse effects that were raised by Young and Kochert and Nietfeld *et al.*: Does the information obtained from the study justify marking of animals?

Can the effects of marking be identified during data analysis? If marking effects are accounted for in the analysis, can the study objectives still be achieved? Such questions should be posed at the outset of any study involving the marking of animals. If one or more answers to these questions is negative or unknown, an alternative marker should be sought or the effects of the marker under consideration should be evaluated thoroughly.

EFFECTS OF MARKERS AMONG TAXA

We reviewed a sample of articles that had as a primary objective the evaluation of marker effects. The articles consisted of qualitative or quantitative assessments of the effect of specific marker types on study animals. We acknowledge that marker evaluation studies probably are biased towards those showing effects because results indicating no effects might be published less often. This implies that our sample of the literature overestimates the occurrence of marker effects in evaluation studies. However, the objective of our review is not to determine how often marker effects occur, but rather to provide examples of the range and diversity of negative effects among marker types, species, and sex, and thus encourage biologists to consider seriously the effects of marking animals. Our review begins with these examples, presented by taxonomic group in the following sections and associated tables.

FISH

Tagging

Marking has been used widely in fish population estimation; accordingly, the earliest tests evaluating marker effects in vertebrates occurred in fish. Historically, most evaluations of marking effects were anecdotal, but by the 1940s researchers were suspicious of the potential effects of markers and thus began evaluating their merit in the field. Early fish research often involved the use of commercially made plastic or metal tags, and fish tagging was considered an effective marking system because tags were inexpensive, easily applied and seen, and rarely lost by tagged fish. However, studies evaluating potential effects of tags often found that tags altered aspects of fish biology. For example, several field studies used mark—recapture techniques and concluded that tags reduced survival and growth of fish. In some situations, negative effects persisted throughout the life of a fish, whereas in others, the effects seemed to be short-lived. Tagged fish were found to experience reduced swimming ability because of increasing drag, but not all effects of tagging can be attributed directly to the tags themselves.

For instance, choice of tag placement on the fish's body can elicit marker effects, and it is generally considered that tags placed in and around the mouth may interfere with feeding. It is notable that not all tag evaluation studies have shown negative effects of tagging, and with additional study some tags will be shown to be more appropriate than others. The recent development of passive integrated transponder (PIT) tags has allowed researchers to mark fish and other vertebrates with smaller tags than those used previously. PIT tags are electromagnetically charged microchips implanted either subcutaneously or intraabdominally, and are read remotely via a portable scanner. Insertion of PIT tags usually is performed using a syringe, thus eliminating the need for extensive invasive surgery. So far, no negative effect of PIT tags has been found in fish, suggesting that this technique can become an important tool for marking fish and other vertebrates. However, one drawback of PIT tagging is that at present tags can be read only when near a scanner.

Mutilation

Marking by mutilation, usually by fin removal or partial removal, is a permanent marking technique often used by fish researchers. However, fin removal often affects fish growth and survival. For example, evaluations using mark—recapture methods show that fin removal causes lower probability of recapture and, by inference, lower survival. Excision of multiple fins generally appears to be more harmful than single-fin excision, and removal of the adipose fin usually is less harmful than removal of other fins. Removal of dorsal or anal fins can be particularly damaging, partly because under certain conditions such excisions may predispose some species to bacterial or fungal infections or predation. However, as with tagging, not all studies evaluating fin removal have detected significant effects, suggesting that for certain species or age classes, or under specific conditions, this marking technique could be acceptable. Clearly, evaluation of the effects of fin clipping on fish biology requires more attention, particularly under controlled laboratory conditions.

Radiotransmitters

Radiotelemetry has become an important technique in fishery research, allowing biologists to accurately monitor long-term movements and survival of many species that would otherwise be difficult to study. Transmitter sizes and types available for fish are variable, and they have been attached to animals either externally or internally. Laboratory studies have shown that externally mounted transmitters increase drag and reduce or prevent swimming, particularly in high-speed currents. It has been suggested that fusiform, lotic fishes are more influenced by external mounts than nonfusiform, lentic, or pelagic species. Internal implantation can be achieved by force-feeding stomach transmitters, or by surgery to attach the transmitter either in the peritoneal cavity or intramuscularly. Implants are more commonly used than external transmitters and have the advantages of lying near a fish's centre of gravity, not being lost or entangled in the environment, and not creating drag forces.

However, these advantages can be offset by reduction in swimming performance, increased handling time, and stress associated with surgery, as well as the higher chance of infection following release. Also, implanted transmitters occasionally can be passively expelled from the body, although sometimes without causing mortality or morbidity. Some species appear more predisposed than others to postoperative complications and transmitter expulsion, meaning that it may be necessary to tailor surgical technique and specific implantation site to the target species. However, in some species, stomach implants seem to have fewer effects than either external mounts or surgically implanted transmitters.

In all telemetry studies, transmitter size is an important consideration, and smaller transmitters are always more desirable than larger ones from the standpoint of effects on the animal. However, the general question regarding the effects of transmitter mass on fish still must be addressed in controlled studies.

REPTILES AND AMPHIBIANS

Tagging

The use of marking in reptile and amphibian research is fairly new, so fewer studies have evaluated marker effects in these taxonomic groups. Many species of reptiles and amphibians have proven difficult to mark because of their epidermal sensitivity, small size, and potential for tissue regeneration. Tagging of reptiles and amphibians has included various types of branding and the use of polymers, pigments, dyes, and radioactive substances. Many of these markers are of limited utility because they were not tested adequately for marking effects; such limitations are particularly important for amphibians, given the sensitivity of their skin. A field test of marking by dye injection did not find any effects on larval amphibians, but a controlled laboratory study did identify stunting in dyed tadpoles.

Although these studies used different dyes, the results call into question previous suggestions that some dyes are largely benign and suggest that laboratory studies might be more sensitive to detection of marking effects. Other colour markers, such as fluorescent paint, often are used to monitor amphibians in the field, despite the fact that such paint apparently has not been evaluated for negative effects on behaviour, physiology, or vulnerability to predation. Most metal or plastic tags used on reptiles and amphibians were originally designed for attachment to fish or birds. Such tags tend to be large and cumbersome, and their effect on study animals remains largely untested despite early suspicions that they affected behaviour and physical condition. It is promising that both studies evaluating the effect of PIT tags on reptiles failed to detect effects, and with additional study, they might become the standard for tagging many species of reptiles and amphibians. However, in addition to expense and distance requirements for reading, PIT tags have the disadvantage of being lost at a high rate by some free-ranging reptiles.

Mutilation

Until recently, most studies of reptiles and amphibians used mutilation marking to identify individuals. One of the most common forms of mutilation, toe clipping, has been widely used on lizards, frogs, and salamanders because it provides an inexpensive method of identifying individuals. High frequency of natural toe loss in some populations of free-ranging lizards has been used to justify its use as an acceptable marking tool, but the natural occurrence of missing toes does not indicate that toe loss is not traumatic. Although toe clipping apparently does not affect the sprint performance of some lizards, another study inferred from the low rate of recapture of toe-clipped toads that the marking technique reduced survival. Clarke also noted that recapture rates were inversely related to the number of toes removed, and that toe-clipped toads experienced reduced dexterity when handling large prey. In other species, regeneration of clipped toes can occur, thus causing problems associated with misidentification of marked animals. However, despite the potential negative effects of toe clipping on reptiles and amphibians, it has remained a widely used form of marking. Clearly, the effect of this technique on reptiles and amphibians requires additional study.

Radiotransmitters

Movements of reptiles and amphibians occasionally have been monitored using a thread-loaded bobbin that unrolls a trail of thread as the animal travels. More often, however, radiotelemetry is used to monitor movements, behaviour, and physiology of reptiles and occasionally amphibians. For amphibians, the main constraint appears to be related to transmitter size, and as a general rule it is recommended that packages not exceed 10 per cent of the body mass of the study animal. Radiotelemetry is problematic with many reptiles and amphibians, and snakes in particular offer challenges because external mounting is not feasible.

Considerable effort has been invested in developing an effective method for implanting transmitters in snakes. However, the value of stomach implants is questioned on the grounds that they may affect aspects of snake behaviour. For instance, stomach-implanted snakes seem to behave similarly to nonimplanted snakes that have recently ingested food, suggesting that activity patterns of implanted snakes are not representative of those of nonimplanted animals. Stomach transmitters also can affect other behaviours or physiological processes, and it might be that such markers simply are not acceptable in snakes because of effects on the animal. Alternatively, transmitters can be implanted in snakes either intraperitoneally or subcutaneously, and these modes of attachment generally appear to be effective.

BIRDS

There is more literature about the effects of marking on birds than for other taxa. Therefore, we provide a sample of recent references for numerous avian marking techniques, and refer the reader to recent reviews by Nietfeld *et al.* and Calvo and Furness for earlier references.

Bands And Collars

Selecting correct band size and material is a very important step in the marking process because different band materials and configurations can have different effects on birds.

For example, aluminum butt-end and lock-on bands can cause more injury and reduce the probability of recovery or recapture of some birds, compared to stainless steel. Some authors recommend use of incoloy (a metal alloy) bands as a substitute for aluminum bands because aluminum bands can cause abrasion to legs of some bird species. Young birds whose legs are still growing can be the most subject to harmful effects of improperly fitting bands, but one method alleviating such effects is by using plasticine to fill the space between standard size bands and the leg. With this innovation, young birds can be banded without the risk of the large band injuring the bird or slipping off the leg.

Band colour can influence bird behaviour. Burley noted that for captive zebra finches (Poephila guttata), interactions among the sexes and mortality were influenced by legband colour. Among wild zebra finches, Zann found no differences in survival or body condition associated with legband colour, but in one colony females that paired with red-banded males laid more eggs than females paired with males not banded with red. These results and others reveal that the effects of colour banding can be complex because they vary by species and experimental and environmental conditions. The conspicuousness of colour bands has been enhanced by attaching streamers of the same colour to the band. However, the durability of streamers and fading as well as birds attempting to remove the streamers are problems.

Colour marking also has been accomplished by placing coloured tape that contrasts with the plumage on several adjacent flight or tail feathers. Also, feathers can be dyed or painted to enhance detectability, or a portion of coloured feather can be used to replace a natural feather. However, like colour banding, colour marking of feathers can cause either significant or negligible effects, depending on various conditions. Neckbands, which are similar to legbands, have been used on long-necked bird species because they are more easily seen and read. In some cases, neckbands have been found to affect bird survival. Patagial tags, also known as wing tags, are used to enable identification of individual waterfowl.

However, in a study of American coots (Fulica americana), patagial tags were associated with loss of body mass when compared with neckbanded controls. In other species, the use of wing tags may result in wounding, changes in migration times, and reduced reproductive success. There are other types of bird markers, many having been shown to have effects on birds. For example, titanium dioxide, which was a useful marker on some species, was found to be deleterious to several raptor species. Fluorescent bead markers can be spread in water, from which they attach to waterfowl, and apparently cause no irritation or detectable physical change in the birds. However, the overall efficacy of the marker is subject to exposure time, bird activity, marker transfer among birds, and equipment required for recognition of marked animals.

Radiotransmitters

Radiotelemetry is an important tool for the study of avian biology. Radiomarking and use of recording devices (*e.g.*, for flight velocity, diving depth) on birds has generated considerable study of effects of these markers because they are large compared to most other bird markers. The effect of transmitter or device size on birds can be influenced by where the package is placed or how it is attached. Packages have been placed on legs, necks, wings, backs, retrices and other feathers, under the skin, in the body, by banding, collaring, wing tagging, harnessing, gluing, tying, suturing, clamping, and implanting. Since the earliest uses of radiomarking in birds, it has been recognized that the transmitter attachment method can affect a variety of aspects of behaviour and survival.

For instance, neck collars were shown to be effective in some cases, but in other cases their use was accompanied by negative effects. In one series of studies, tail-mounted transmitters did not affect mass or survival of northern goshawks, but it is understood that such transmitters must be light (2 per cent or less of bird body mass), thereby limiting battery size and transmitter longevity. Reid *et al.* describe a method for replacing a tailmounted transmitter when individuals can be recaptured. As with legbands and dyes, transmitter colour also must be considered: Wilson and Wilson and Wilson *et al.* V found that penguins pecked significantly less at black recorders (attached to the dorsal feathers by tape) than at other colours. Radiotransmitters are similar to other markers in having variable effects that are influenced by attachment method, and the bird's species, age, and sex are necessary considerations.

In an effort to securely attach large packages to birds, a number of researchers have experimented with various harness designs to hold transmitters on the bird's back. Several authors failed to show differences in survival or behaviour of birds carrying various harness transmitter packages, but other examples show that such transmitters can have negative effects on bird behaviour or survival. In an attempt to minimize the deleterious effects of harnesses on birds, researchers have experimented with different types of implants. Partialimplant as well as full-implant transmitters have been used with some success, but such implants can result in short-term preening over the incision site and cause low rates of seroma and infection. One constraint with transmitter implants is that transmission distance is reduced when the antenna is implanted in the abdominal cavity, although some researchers have developed a technique to exit the antenna from the body, thereby augmenting transmission distance.

Since the mid-1980s several investigators have considered the effects of radiomarking on bird energetics. This issue is important for birds because radiomarking compounds the increase in energy required to carry additional mass by adding aerodynamic or hydrodynamic drag. Streamlining of transmitters reduces aerodynamic drag and therefore minimizes their negative effects. Although some researchers failed to detect effects of transmitters weighing less than 4 per cent of body mass, others indicate that, either for different species or larger transmitters, radiomarking can affect bird metabolism. This may be particularly important for large birds because they have proportionally less surplus power than smaller birds.

MAMMALS

Tagging

Although studies of mammals often involve marking,

marker effects have been evaluated in few instances. This appears to be particularly true for externally mounted metal or plastic tags, with few studies evaluating effects of such markers despite their widespread use. Internal PIT tags have been used in several species of mammals and evaluation tests so far have failed to detect significant negative effects.

Mutilation

Toe clipping is a widely used tool for marking small mammals, and many studies have evaluated the effects of this technique on survival and body condition. Several studies have detected effects of toe clipping, but similar numbers of studies have failed to observe negative effects. In one case, different studies on the same species provided conflicting results, suggesting that study methodology can influence outcome of marker evaluation studies. In two cases, significant effects of toe clipping were restricted to a particular sex, implying that not all animals are equally vulnerable to negative effects. However, the prevalence of toe clipping as a tool for marking small mammals necessitates that additional studies address its potential effects. Other forms of mutilation include freeze branding, tattooing, and fur clipping, but in most cases effects of these markers have not been evaluated. However, Stewart and Macdonald did show that European badgers (Meles meles) could be effectively marked via fur clipping and that the mark has no effect on badger body condition. However, the applicability of this technique in colder climates, where loss of guard hairs may affect thermoregulation, requires further study.

Radiotransmitters

Radiotelemetry of mammals can involve either external or internal attachment of packages. We found that small mammals had received more attention than other groups in marker evaluation studies and that only recently had terrestrial carnivores received consideration for potential transmitter effects. Most studies we surveyed failed to find significant effects of transmitters, but discrepancies among studies performed on the same species were noted. For instance, the meadow vole (Microtus pennsylvanicus) was subjected to four evaluation tests:

Bertaux *et al.* did not find a negative effect of transmitters (6.7 to 9.0 per cent of body weight) on vole energetics, but other studies showed that transmitters affected vole activity patterns. The fact that differential activity of radiomarked voles was not detected as higher energy expenditure highlights the difficulty associated with attempting to generalize study results. It also has been shown that effects of radiotransmitters on small mammal behaviour and movements often are either short-term or specific to a particular sex, but the potential demographic implications of such marker effects have not been assessed.

CRITIQUE OF MARKER EVALUATION STUDIES

A primary shortcoming of many marker evaluation studies is experimental design. Sometimes this is manifested as a lack of appropriate controls (*i.e.,* unmarked animals) to which marked animals can be readily compared. The lack of control animals was common during the initial period of marker evaluation studies, but more recent evaluation studies also possess this flaw. In other cases, authors include in the design a cohort of previously marked or alternatively marked animals as controls. However, this approach assumes that "control" animals are representative of the unmarked population even though alternative markers might cause important effects on their own. In this case, any comparison of marked versus "control" animals could result in an underestimation of effects of the targeted marker.

In some studies, control animals are not subjected to the same handling procedure as marked animals, thereby making marking effects indistinguishable from those of handling. This can be particularly problematic in situations where handling causes significant stress or long-term effects, and researchers may find it difficult to identify which procedure (marking or handling) requires modification. However, some studies have correctly subjected controls to all the same handling procedures as the marked sample, thus allowing a more rigorous evaluation of the effects of the marker itself. Marker evaluation studies often have sample sizes that are simply too small to detect a reasonable difference between marked and unmarked samples. Inadequate statistical power increases the likelihood of committing a type II error, thereby increasing the chance of failing to reject a null hypothesis of no significant marking effects when effects actually occur.

Marker effects tend to be more readily detected in the laboratory because field studies often have smaller sample sizes and larger within-sample variance. Many field studies for which marker evaluation is apparently an offshoot, or those that evaluate marker effects on large mammals, lack statistical power. Thus determining the detectable effect size and statistical power associated with a given marker evaluation study should be a necessary precursor to implementation of that study. Also, whenever possible, studies probably should be initiated under controlled laboratory conditions to reduce confounding effects of the environment. However, laboratory studies should be followed by evaluations in the field. A common characteristic of marker evaluation studies is the use of subjective or qualitative measures of marking effects. Without rigorous statistical treatment of measured effects, results of such studies are of limited utility. Also, indices are sometimes used to infer direct effects (*e.g.*, calculating capture—recapture rates to infer marking effects on survival), but if the index also measures other aspects of species biology (*e.g.*, dispersal), such inferences might be spurious.

Other evaluation studies are too short to derive meaningful conclusions regarding long-term effects, even though the latter effects may very well be the most demographically significant. Finally, sometimes statistically significant results are not considered to be biologically important because they are too small or uncommon. Each of these approaches reduces the likelihood of identifying marker effects that may adversely affect the animal or the study results. Our review of the marker evaluation literature reveals that a marker can affect a variety of aspects of animal biology, and that different types of evaluations provide different results. When biologists plan an evaluation of marker effects or when they interpret and apply results from previous evaluations, they must make decisions about which methods of evaluation are most appropriate for their objectives and subject species. Also, they must decide which results are most relevant to assessing the importance of an effect on the animals and their study objectives. For example, a biomechanical analysis of the effect of a marker provides an estimate of how much extra energy is needed to carry the marker, but it does not determine whether that increase in energy expenditure has other biological implications for the animal, such as reduced food delivery to young. A metabolic measurement might not indicate a significant change in O_2 or CO_2 between the marked and unmarked animals. Should we thus conclude there is no effect? Observations of behaviour of the same animal might reveal that the marked animals spend more time resting than unmarked animals. All these evaluations could produce "significant" results, and yet contradict each other or provide different types of information. It is the responsibility of the researcher to conduct or consider the most appropriate marker evaluations relevant to the study objectives and the well-being of the study animals. The biologist must decide which effects are important.

CURRENT GUIDANCE AVAILABLE FOR CHOOSING MARKERS

Numerous criteria must be considered when selecting markers to be used in a given ecological study, including potential effects of markers on animals. Researchers can review the literature, refer to colleagues, or consult with marker manufacturers or merchants. Limited guidance also is available through guidelines published by scientific journals and government agencies and in general references. In addition, several professional zoological societies have published guidelines for the use of animals in field research. In general, these societies suggest that markers incur as little pain as possible and not restrict excessively behaviour, physiology, and survival of study animals. We provide a brief overview of some current recommendations provided by zoological societies

TAGGING AND MUTILATION

Tags used on fish, reptiles and amphibians should be of appropriate size and shape, but the use of tags that protrude from the body or are brightly coloured is discouraged. It is recommended that for birds all bands be of appropriate size, but the use of nasal disks, saddles, patagial markers, dyes, and ultraviolet markers is discouraged. Fin clipping is suggested as having minimal impact on survival and social structure of fish, and it is recommended as an appropriate technique if the specific fins to be clipped are expendable by the target species. Free-ranging reptiles and amphibians should not be toe-clipped unless the technique has been shown not to impair normal activity in the target species or a close relative, whereas for mammals it is recommended that all types of mutilations be avoided.

Although birds are occasionally marked via mutilation (*i.e.*, nail clipping, web-punching, feather clipping), no guidelines for use of this technique have been provided. Marking fish, reptiles, and amphibians using techniques such as tissue removal, branding, freeze branding, and electrocauterization is generally acceptable, but the use of tattoos and paint is less desirable because of problems associated with dye visibility and legibility. Although fish can be marked with paint, the technique is discouraged for use on amphibian skin, for which nontoxic stains and dyes should be used. In cases where toxicity is unknown, laboratory trials should be undertaken before any field use. Few guidelines are provided for the use of brands, dyes, or paints with birds and mammals.

Radiotransmitters

Most professional zoological societies address the issue of radiotransmitters specifically. Many fish, reptile, and amphibian species are not suitable for radiotelemetry because of their small size. However, for species that are amenable to telemetry, stomach implants and internally mounted transmitters should be small and coated with a biologically inert coating, and not interfere with physiology and behaviour. Externally mounted transmitters should be shaped and attached to reduce chances of entanglement, irritation, or constriction.

In the case of large birds, it is suggested that radiotransmitters weigh less than 1 per cent of body mass to reduce negative effects on biomechanical performance. For smaller birds transmitters should not exceed 5 or 10 per cent of body mass. Before use in the field, biologists should observe individuals in captivity to evaluate effects of radiomarking on behaviour. For reptiles, amphibians, and most mammals, it is recommended that transmitters not exceed 10 per cent of body mass.

CRITIQUE OF GUIDELINES AVAILABLE FOR CHOOSING MARKERS

In general, the guidelines provided by zoological societies are too general for choosing a specific marker for a given study objective or species. Some recommendations made by professional societies even appear to ignore the findings of previous marker evaluation studies. For instance, fin removal is recommended by several fish societies as an appropriate method of marking many species, despite numerous instances in which the technique has been shown to affect fish biology. Also, it is recommended that mass of transmitters never exceed 10 per cent of body mass of vertebrates, even though some transmitters weighing less than this have been shown to produce negative effects.

Given that, at least for birds, the effect on flight power increases with body mass, the 10 per cent threshold is clearly an arbitrary construct that does not apply to all species. Furthermore, a given mass or drag has different effects depending on the type of flight. Therefore, general guidelines (*e.g.*, 10 per cent, 5 per cent, or 3 per cent) can be misleading.

Although for terrestrial animals costs associated with carrying a load increase linearly with load mass, intra- and interspecific differences in biology should preclude the establishment of a threshold that transcends taxonomic groups. Societies should assume a more comprehensive responsibility to indicate to their members possible shortcomings associated with various marking procedures, although it remains the responsibility of individual researchers to address the potential effects of markers on their study species. Where information is lacking on a particular marker or its effect, zoological societies should encourage additional evaluation studies. It is imperative that markers used in research be acceptable to animal care and use committees before project initiation, and zoological societies should strive to harmonize their standards. With those set by such committees and disseminate information regarding acceptable markers and protocols to committees. Otherwise, research using newly tested markers might be needlessly precluded because of an uncertainty regarding marker effects.

RECENT ECOLOGICAL STUDIES

In light of the measurable and obvious effects of markers to numerous vertebrate species, we examined the treatment of potential marking effects in a sample of recent peer-reviewed literature. Specifically, we assessed the frequency with which researchers addressed potential marker effects, either by using methodologies that reduced potential effects or by testing for such effects qualitatively or statistically. We surveyed nine journals that publish studies on a broad range of taxonomic groups for articles in which vertebrates had been marked; our survey included only articles in which the primary objective was to address general issues related to animal biology rather than to study marking effects.

We found that in most instances (90 per cent, n = 238), authors did not address potential effects of marking, or at least did not report any such consideration in the article. Undoubtedly, some authors attempted to minimize or evaluate marker effects but chose not to report these efforts in the article;

however, we suspect that such cases were uncommon. We found no indication that the failure to evaluate marker effects was weighted towards a given taxonomic group or journal. In 3 per cent of cases authors assumed explicitly that markers did not affect animals.

Such assumptions were usually based on previous reports showing no significant marking effects in the target (or related) species or on requirements of the statistical method chosen to analyse the data. Only 7 per cent of the articles surveyed included information directly pertaining to potential marking effects. Some authors attempted to minimize marker effects by allowing postmarking recovery to take place in captivity. Although this approach might be effective when handling and marking are stressful or invasive (*i.e.*, internally mounted radiotransmitters), it might also bias study results if captivity affects behaviour or survival after release. For species that are highly vulnerable to stress from captivity, prolonged recovery under controlled situations could be less desirable than immediate release.

Other attempts to minimize marker effects included adjusting markers to fit individual animals or using expandable and breakaway markers. Some authors justified the use of a marker by describing qualitatively how animals appeared to behave normally after marking. Subjective evaluation of potential marking effects is a good first step, but whenever possible quantitative measurements of the biology of marked versus unmarked animals or comparisons of preand postmarking behaviour of individuals should be included. When markers were suspected of exerting short-term effects on animals, researchers excluded data obtained during an arbitrary period.

However, this practice can result in data bias if marking causes subsequent increases in mortality of young or frail animals; the result would be an inflation of survival estimates. Two studies evaluated the effect of one marker type using an alternatively marked cohort as control, without validating that negative effects of the alternative marker did not occur. We found no indication from our survey that authors addressed potential long-term effects of markers, although one study inferred from the lack of external damage in study animals that long-term tagging mortality was absent or negligible. In another case marked and unmarked animals were monitored and data from both cohorts were apparently pooled, but no mention was made of trying to evaluate possible marker effects. It is also notable that almost all (more than 90 per cent, n = 16) instances in which authors addressed marking effects the studies involved animals that were marked with radiotransmitters.

This implies that various forms of mutilation and tagging or banding are either assumed by researchers to have little or no effect or else not afforded the same scrutiny in the editorial process. To summarize, our review of ecological studies found that in general authors tended to overlook or not acknowledge potential marking effects.

In few cases potential marking effects were addressed, but usually only in a superficial manner or one that could produce bias in the resulting data. This review clearly indicates that there are a large number of reported marking effects from commonly used markers, there is a lack of thorough guidance from zoological societies in choosing a marker, and there is a common failure to address potential marking effects in the ecological literature. Thus, we provide the following recommendations for using markers and for additional research into the effects of marking wildlife.

PROTOCOLS AND TECHNOLOGICAL ADVANCES

Depending on study objectives, target species, and possible problems associated with the handling and marking process, it is not always necessary or desirable to apply marks to animals. For example, in some studies genetic or mineral markers in animal tissues, excreta, or blood, may allow for identification of individuals or populations. However, reliable identification using genetics is, at present, labour intensive and costly, can require recapture, and is largely untested under field conditions. As another alternative, it might be possible to use naturally occurring variable colour markings or unique morphological features for recognition of individuals. For example, some amphibian species are amenable to visual recognition of individuals, and photographs or sketches can be used to record characteristics of individuals in a population. Individual recognition via natural markings also has been used occasionally for fish and some bird species, but has probably found its greatest utility with large mammals.

For mammals, researchers usually rely on unique facial features to identify individuals; in most cases the method appears reliable although validation is desirable and there are limits to the number of animals in a population that are individually recognizable. Pennycuick reviewed limitations and difficulties associated with the use of natural markings in free-ranging animals; where the use of this technique meets study objectives, it is more desirable than artificial marking. It is essential that field personnel be adequately trained in the proper handling and marking of animals. This may involve practice with captive individuals of the targeted or surrogate species, or animal models.

Improper marker application by poorly trained field personnel can result in serious problems for study animals, and well-trained personnel not only minimize the potential for negative marker effects but also are able to process and release study animals more rapidly, thereby reducing potential negative effects of handling and marking. Although significant progress has been made in the development of less harmful marks, additional steps are necessary. Few technological developments can contribute to minimizing effects of mutilations, although routine use of surgical instruments and sterile conditions lessens occurrence of disease and mortality.

Development of smaller tags and bands can contribute to less invasive marking. In the past, one constraint of reducing tag size was the ability to read numbers or colour markings identifying individual animals. However, carefully designed protocols can increase. The use of PIT tags eliminates size constraints for many animals. However, other tags also hold promise, such as passively applied (*i.e.*, usually ingested) group recognition markers. These may prove particularly useful when animal handling is undesirable and individual recognition is not necessary. Radiotransmitters probably will remain an important and widely used form of marking, and technological developments should include aerodynamically or hydrodynamically shaped, smaller, and lighter packages and development of new attachment methods.

Because most of the bulk caused by transmitters is attributable to batteries, packages could be significantly lightened if smaller high-capacity batteries were developed. However, in some cases significant reductions in transmitter size or mass are not possible without forgoing battery longevity. Thorough testing and reporting of animal marking procedures provides other investigators information useful for choosing among various attachments and trying them with a new species or under new circumstances. Detailed descriptions of procedures and modifications to procedures provide potential users with additional options.

Finally, the results of applying various attachments in field situations are very important and should include information about effects of the capture and handling that must accompany marking. Sometimes the procedures associated with capture and handling can be as important a component in the marking process as the marker itself. Our review reveals such a variety of marker effects (depending on marker type, species, age, sex, attachment method, environment, and season) that careful development, testing, documentation, and reporting of experience with markers will continue to be very useful in identifying and dealing with marker effects. Zoological societies should assume the responsibility for setting reasonable standards for the marking of animals by providing detailed guidelines.

MARKER EVALUATION STUDIES

We recommend increased efforts to test for effects of markers, changes in existing protocols and recommendations, and more publication of the results of marker evaluation studies. Evaluation studies are a critical step in the marker validation process, but many commonly used markers have been evaluated for only a few species. Our review highlighted several specific deficiencies in our understanding of marker effects, such as effects of radiotransmitters in fish, efficacy of internally versus externally mounted transmitters (particularly stomach implants), influence of external tags on amphibians and mammals, and effects of transmitters on large mammals, particularly carnivores.

It will not be possible to evaluate the effects of all markers in all species, but results of evaluation studies will be more useful if researchers focus their initial evaluation studies on broadly applicable markers and species that share several biological features with other species of interest. We think that PIT tags will become very important for marking animals in the future, and thus urge researchers to undertake the necessary evaluation studies without delay. All evaluation studies should be designed to measure marker effects across ages and sexes.

Evaluation studies must include comparison to unmarked (or alternatively marked) controls and adequate sample sizes. Statistical power analysis can be used to calculate the minimum sample size required to detect significant effects. Captive or laboratory conditions should be used whenever possible, although researchers should ensure that biological parameters being measured are not significantly affected by captivity. Ultimately, results of captive studies should be validated in the field. Also, evaluation studies should take advantage of sensitive techniques for estimating subtle negative effects, such as biomechanical modeling and doubly labeled water. Marker evaluation studies should attempt to address indirect effects of markers on survival (*e.g.*, predation, starvation, disease) and behaviour (*e.g.*, altered foraging or parental abandonment of juveniles to compensate for higher energy demands), and focus not only on short-term but also on long-term effects of markers.

Marker evaluation studies should be published in the peer-reviewed literature. It was not possible for us to determine whether a publication bias existed in our sample of marker evaluation studies, but we suspect a tendency exists for publishing studies that show significant negative effects over those failing to show effects. Such a bias can misrepresent the effect of markers, and it inhibits dissemination of information about the best marking methods. Clearly, some journals must encourage authors to perform and publish marker evaluation studies.

A similar suggestion was proposed to overcome the apparent publication bias in biological studies showing nonsignificant results. Publication of all marker evaluation studies would provide researchers and animal care and use committees with current knowledge regarding acceptable animal markers. It remains the responsibility of reviewers and editors to ensure that appropriate marker evaluation studies are cited or conducted in the published ecological and wildlife literature. In some cases, it may be sufficient to justify the use of a given marker by citing previous studies that have failed to detect effects, but when researchers are applying a marker to a new (or unrelated) species, or using a marker that has not been evaluated previously, appropriate tests must be undertaken. Again, experimental design and sample size are paramount considerations for tests. Publication of technique validation studies is common in other scientific fields and should be encouraged in animal ecology.

Because current funding sources often do not support technique evaluation studies, researchers must find other sources of funding for such work, such as private animal welfare organizations, and they must encourage traditional sources of funding to support evaluation of marking methods. Only through additional studies and published results will it be possible for colleagues to thoroughly evaluate potential marker effects and choose the most appropriate markers.

8

Methods for Evaluating Habitat Selection

HABITATS

The first prerequisite for assessing habitat selection is that habitats be defined as discrete entities. For use—availability studies in particular, the defined number of habitats can directly affect the results. Yet habitat distinctions often are not clear-cut. A researcher might distinguish two general forest types, uplands and lowlands, or might classify habitats by dominant overstory, or might divide these further by stand age or understory, and so on. As more types are defined, sample sizes are reduced for observed use of each type, thereby diminishing the power of the statistical tests to distinguish differences between use and availability. Also, because the proportional use and availability of all habitats each sum to 1, the number of habitats distinguished affects all of these proportions. Aebischer *et al.* observed that this unit—sum constraint renders invalid many of the statistical tests often employed to compare use and availability because the proportions are not independent. That is, if one habitat type.

Has a low proportional use, others will have a correspondingly high use, and if there are only a few types, then the infrequent use of one type will lead to the apparent selection for another. Aebischer *et al.*'s method of compositional analysis was developed specifically to circumvent this problem. Not just the number of types, but the criteria used to partition types may greatly affect results.

Knight and Morris were able to visually differentiate 13 habitat types on landscape photographs of their study area, but postulated that only two broad classifications were distinguished by red-backed voles (Clethrionomys gapperi), the subject of their study. After analysis of their data, however, it became clear that from the voles' perspective, at least three functional habitats existed.

Another problem is the scale at which habitats are viewed. For example, an animal might appear to select for a certain habitat type, defined by a dominant cover type, whereas in reality it selected for certain specific kinds of sites that just happened to occur more commonly in that cover type than in others. An animal's choice of habitat type is often called macrohabitat selection and the choice of specific sites or patches within habitats is called microhabitat selection. These may be perfectly hierarchical in that the most preferred microhabitats always occur within the same macrohabitat, in which case an animal may really select initially at the scale of macrohabitat, and then focus on specific sites within it. Schaefer and Messier observed this sort of nested hierarchy across a range of scales for foraging muskoxen (Ovibos moschatus) in the Canadian High Arctic.

Alternatively, the distribution of preferred microhabitats could be largely unrelated to the broader habitats defined by the biologist; in this case, a site attribute study might identify characteristics related to preferred microhabitats, whereas a use—availability study would detect no selection at the level of habitat type. This situation was apparently the case for wood mice (Apodemus sylvaticus) inhabiting arable lands in Great Britain: The mice seemed not to select (based on a use—availability study) from among three types of croplands (macrohabitats), but within each of these croplands they chose microhabitats with a high abundance of certain plants.

In sum, significant challenges in defining habitats include: partitioning them in terms of the features that the animals are selecting for, which are not necessarily the ones we most easily discern; delineating sufficient habitat categories to ensure that the truly important types are not lumped with and thus diluted by less important types; and not diminishing the power to discern selection by parceling out too many types.

MEASURING HABITAT USE

Sample bias is an obvious potential problem in measuring habitat use. Interpretations of habitat use from visual observations of animals or their sign can vary among observers and sightability can vary among types of habitats, both of which can introduce biases in the data. For example, Powell noted that fisher (Martes pennanti) tracks in snow were difficult to follow in habitats with dense vegetation, especially where fishers followed trails of snowshoe hares (Lepus americanus); in this case the bias against observing tracks in dense vegetation merely detracted from the overall conclusion that densely vegetated habitats were frequently used. Counts of pellet groups (*e.g.*, from ungulates or lagomorphs) may poorly reflect habitat use because defecation rates often vary with the food source, and hence the habitat type. Capture locations may be a poor indicator of habitat use because baits and other trap odours (*e.g.*, from captures of other animals) may affect behaviours in an unpredictable way.

Telemetry also may yield biased data on habitat use because the detection of an animal's radio signal may depend on the habitat it is in, and location data obtained by triangulation have inherent associated errors. Intuitively, and as shown in computer simulations by White and Garrott errors in determining habitat use increase with increased habitat complexity and decreased precision in the telemetry system.

Errors do not necessarily introduce bias, but can if patch size differs among habitats (detected use would be underrepresented in habitat types that tend to occur as small patches) or if the animal preferentially used the edge of some habitat types but not others. Powell reported different perceptions of habitat use of fishers between his study, where he followed tracks in the snow, and another nearby radiotelemetry study; he attributed the difference to error in the telemetry system and consequent incorrect habitat categorization for animals near edges.

Nams showed that simply discarding locations because of large telemetry error, as is common practice, exacerbates bias; he offered a procedure for circumventing it, but few studies have used it. Kufeld *et al.* suggested using the habitat composition of error polygons formed by the triangulation of radio bearings, but this would not alleviate bias. Chapin *et al.* solved the problem a different way. In a study of habitat use of American martens (Martes americanus), which have a documented affinity for mid- to late-successional forests, they classified telemetry locations that were outside small patches of forest but within telemetry error of the edge as representing use of those forest patches. Even if habitat use can be measured accurately, biases may result from sampling or analytical procedures. Habitat use may vary by individual, sex—age group, social status, time of day, season, and year, yet many (most) studies pool individuals and do not sample adequately. Schooley reviewed habitat studies published in the Journal of Wildlife Management and found that most lasted only 2 years, and most pooled results across years without testing for annual variation.

He used results of a black bear (Ursus americanus) study to show that habitat use can vary annually, and that the data pooled across years can yield misleading results. Beyer and Haufler found that most published studies that they reviewed collected data only during daylight hours; in their study of elk (Cervus elaphus), habitat use differed between day and night. Similarly. Arthur and Schwartz reported diurnal and nocturnal differences in habitat use for brown bears (Ursus arctos) that fed at a salmon stream that was used by people during the day; this difference was detected with data from GPS collars, but was not apparent from conventional diurnal telemetry data. Ostfeld *et al.* and Belk *et al.* observed sex-related differences in habitats used by ground-dwelling rodents; Belk *et al.* remarked that combining the two sexes would produce a false perception of habitat use.

Paragi *et al.* observed differences in habitat use of resident and transient martens. Boitani *et al.* and Macdonald and Courtenay observed individual differences in habitat use, apparently related to social status. Bowers found that habitat use of eastern chipmunks (Tamias striatus) varied significantly with distance from their burrows, a finding noticeable only by considering the data on an individual basis. "It is time," Bowers commented, "that ecologists recognize that microhabitat selection and usage is a process involving individuals, not species."

Pooling individuals is common because sample sizes are typically too small to test for selection by individual. However, the statistical tests usually used assume independence among sample units, which is often not the case in studies that consider each location a sample. Some techniques consider animals as sample units, so lack of independence among locations within individuals is not problematic. However, these methods are still subject to difficulties with lack of independence if animals are gregarious or territorial (social exclusion precludes use of certain habitats), or if the study subjects are related (habitat preferences possibly affected by a common learning experience) or are from the same social group. In an effort to alleviate the problem of a lack of independence among individuals, Neu *et al.* used groups of moose (Alces alces) and Schaefer and Messier used herds of muskoxen as their sample units, rather than individual animals.

Similarly, although Gionfriddo and Krausman monitored habitat use of individual radiocollared mountain sheep (Ovis canadensis), they considered groups of sheep their sample unit. However, Millspaugh *et al.* contend that animals in a herd should be considered independent individuals if they congregate because of a resource rather than because of a biological dependence on each other. They provide a hypothetical example with elk, where 99 of 100 radiotagged animals congregated at a winter feeding area in one habitat and the remaining individual used a second habitat; at other times of the year the elk did not associate with each other. In this case, they argue that each radiotagged individual should be considered an independent sample. In contrast, predators that hunt together in a pack and are thus dependent on one another cannot be considered to use habitats independently. Millspaugh *et al.* recommend tests to evaluate independence of habitat use by seemingly associated individuals.

EVALUATING HABITAT

Three general study designs have been used to infer habitat quality. The first, generally called the use—availability design, compares the proportion of time that an animal spends in each available habitat type (generally judged by the number of locations, or less commonly, by the distance traveled) to the relative area of each type. The second, the site attribute design, compares habitat characteristics of sites used by an animal to unused or random sites. These two designs generate measures of selection for various habitats or habitat attributes, and habitat quality or importance is inferred from the magnitude of this apparent selection. The third method, which the demographic response design, uses a more direct approach for assessing habitat quality by comparing the demographics (density, reproduction, or survival) of animals living in different habitats. This design thus circumvents the need to interpret animal behaviour (habitat choices).

STUDIES OF BIRDS AND MAMMALS

Among studies of birds and mammals, the use availability design is the most popular. Habitat-related papers dealing with birds and mammals published in the Journal of Wildlife Management during 1985–1995 and found that most relied on a use—availability study design to assess habitat selection, preference, or quality. Thomas and Taylor further categorized use—availability studies into three approaches: one in which habitat-use data are collected on animals that are not individually recognizable (*e.g.*, visual sightings or sign), one in which data are collected on individuals (*e.g.*, radiocollared animals) but habitat availability is considered the same for all individuals (so individuals are typically pooled for analysis), and one in which use and availability are measured and compared for each individual. They also reviewed papers published in the Journal of Wildlife Management and found that nearly twice as many studies collected data on individuals but pooled them for analysis than either of the other two approaches. Studies that pooled animals for analysis have commonly compared frequencies of use and availability for an array of habitats using a chi-square test.

Determination of which habitat types were used more or less than expected is generally made by comparing availability of each habitat type to Bonferroni confidence intervals around the percentage use of each type. This procedure was described initially by Neu *et al.* and clarified by Byers *et al.*, although a more accurate method of constructing such confidence intervals was recently proposed by Cherry. If the areas of available habitats are estimated (*e.g.*, from sampling) rather than measured (*e.g.*, from a map), use and availability should be compared with the chi-square test for homogeneity rather than goodness-of-fit.

A chi-square goodness-of-fit test assumes that the availabilities are known constants against which use is compared, so if availabilities are actually estimated, with some sampling error, this test is more prone to indicate selection when there is none. Various other methods of comparing use and availability have been advanced but less often used in wildlife habitat studies. Ivlev proposed an electivity index to measure relative selection of food items on a scale from "1 to 1; this has since been adopted for some habitat selection studies. However, Chesson noted that Ivlev's index may yield misleading results because it varies with availability even if preference is unchanged, and advocated use of a 0 to 1 index originally proposed by Manly *et al.*, also for feeding preference studies.

This Manly—Chesson index is simply the proportional use divided by the proportional availability of each habitat, standardized so the values for all habitats sum to 1. As adapted to habitat studies, it is interpretable as the relative expected use of a habitat had all types been equally available (*i.e.*, preference). Thus in an area with four habitats, an index of 0.25 for each habitat would indicate no preference, whereas deviations from this would indicate relative preference for or against certain habitat types. Heisey and Manly *et al.* extended this method to test for differences in habitat preference among individuals or sex—age groups, and also showed how to test for statistically significant differences among preferences for different habitat types.

Kincaid and Bryant and Kincaid *et al.* offered an alternative method that scores relative differences between use and availability for habitats defined as geometric vectors. Most studies using these tests pooled data among individuals, so that animal captures, sightings, radiolocations, and so on represented the sample units. Aebischer *et al.* pointed out that this constitutes pseudoreplication and advised comparing use to availability for each animal individually (*i.e.*, so individuals are the sample units). Several methods have been developed specifically to do this. Of these, the most commonly used is Johnson's, which is based on the difference between the rankings of habitat use and the rankings of habitat availability.

This method also provides a means of detecting statistically significant differences among habitats, not just a relative ordering of their selection. Moreover, because comparisons are made on an individual-animal basis, habitat availability can be considered either within each individual home range, or within the study area as a whole.

Johnson defined first-order selection as that which distinguishes the geographic distribution of a species, second-order selection as that which determines the composition of home ranges within a landscape, and thirdorder selection as the relative use of habitats within a home range. Thus, both second-order and third-order selection can be addressed with Johnson's technique; with chi-square tests it is possible but more difficult (because of sample size constraints) to consider both of these levels of selection. Alldredge and Ratti compared four methods (including the chi-square, Johnson's, and two others based on individual-animal comparisons) in simulated conditions and found that none performed consistently better than the others.

However, some methods are better suited for given situations. For example, because data for all animals are generally pooled for chi-square tests, unequal sampling among individuals could strongly affect the results if all individuals did not make similar selections. Conversely, the methods that weight animals equally, regardless of the amount of data collected on each, may be subject to spurious results caused by small sample sizes and variability among individuals. McClean *et al.* used real data on young turkeys (Meleagris gallopavo), which have fairly narrow and well-known habitat requirements, to compare results of six analytical techniques for assessing habitat selection. In this case, the methods that treat individuals as sample units tended to be less apt to detect habitat selection.

Aebischer *et al.* offered what appears to be an improved procedure for comparing use with availability on an individual animal basis. This method (compositional analysis) has become increasingly popular because it enables assessment of both second-order and third-order selection and yields statistical comparisons (rankings) among habitats. Additionally, because the data are arranged analogous to an ANOVA, in which between-group differences can be tested against within-group variation among individuals, it provides a means of testing for differences among study sites (*e.g.*, with different habitats, different animal density, or different predators or competitors), seasons or years (*e.g.*, with different food conditions), sex—age groups, or groups of animals with different reproductive outputs or different fates.

SITE ATTRIBUTE DESIGN

Site attribute studies differ from use—availability studies in that they measure a multitude of habitat-related variables at specific sites and attempt to identify the variables and the values of those variables that best characterize sites that are used (often for a specific activity).

With this design, the dependent variable is not the amount of use (as with use—availability studies) but simply whether each site was used or unused (or a random location with unknown use); the independent variables can be many and varied. Use—availability studies generally just deal with broad habitat types, or if more variables are considered, they are analysed individually. A site attribute design was used in 45 (29 per cent) of the habitat selection studies. Of these, 28 were on birds and 17 on mammals. This design requires measurement of habitat variables at some defined site, usually one that serves some biological importance to the animal. Nest sites of birds are easily defined and biologically important, and hence are often the subject of studies of this nature. Habitat characteristics of breeding territories, drumming sites, and roosting sites also have been investigated. Among mammals, studies have focused on characteristics of feeding sites, food storage sites, resting sites, shelters, wintering areas, or areas recolonized by an expanding population.

Other studies have compared habitat characteristics of randomly located sites to sites where birds or mammals were observed, radiolocated, or known to have been from remaining sign. The statistical procedures used in such studies vary. Most have used multivariate analyses to differentiate combinations of variables that tend to be associated with the used sites. Discriminant function analysis (DFA) is the most popular of these. Logistic regression is an alternative, and is especially useful when the data consist of both discrete and continuous variables or are related to site occupancy in a nonlinear fashion.

DEMOGRAPHIC RESPONSE DESIGN

Ideally, studies should identify relationships between habitat characteristics and the animal's fitness. Studies employing use—availability and site attribute designs assume that certain habitat features are selected because they improve fitness. Demographic response designs attempt to test this more directly. However, refer to the measured demographic parameters in these studies as response variables, they really only represent correlates with given habitats. Most of these investigated differences in animal density among habitats. Fourteen studies, all on birds, related reproduction (*i.e.*, nesting success) to habitat of nest sites. Three studies, two on birds and one on mammals, attempted to find an association between habitat and survival, but only one detected such a relationship.

MEASURING HABITAT AVAILABILITY

Measuring habitat availability is often more problematic than measuring use. Use—availability studies inherently assume that study animals have free and equal access to all habitats considered to be available. That is, at any given moment each study animal should be able to use any available habitat. This assumption may hold if use and availability are measured for each animal individually. However, the assumption may be violated when animals are pooled for analysis and the available habitat is considered to be the same for all, yet some individuals may not even have all habitat types within their home range.

Johnson suggested that the habitat composition of home ranges compared with the habitat composition of some broader area should indicate the level of selection animals exercise when establishing their home range. Often the broader available area is considered to be that encompassed by the composite of the home ranges of all study animals. However, there are several problems with this.

First, animals cannot really select their ideal mix of habitats to compose their home range. Animals can only choose home range borders that encompass the best mix of habitats from what exists on the landscape; they cannot alter the mix to suit their needs. By analogy, a person may pick a town to live in, among several available, based on the resources available. One may also choose where to live within the town, but one cannot alter the layout of the town or the array of features available.

Second, animals may not have free and equal access to all areas when establishing their home range. Home ranges may be established near the natal area just because of familiarity with resources or neighbouring animals, not any choice related to habitat composition. Analogously, people might remain in their home state or country not because they consciously chose it among all others, but because they never had the opportunity to visit other places, or because moving elsewhere, even if it seemed desirable in some respects, had too many costs.

Social constraints also may dictate choice of a home range by precluding access to certain areas. Extending the analogy with people, consider a house to be like a home range and a neighbourhood a composite home range. The first few residents of a neighbourhood might have selected where to live among houses that differed in various ways; however, as more people moved in, the choices narrowed, until no choice remained for the last resident. If all houses were used, regardless of their quality, one could not discern after the fact which houses were preferred unless the "colonization" process was observed. Fretwell and Lucas proposed a corresponding model for animal populations. In an expanding population, preferred habitats are settled first, but as these are taken, animals are forced to settle in poorer and poorer areas.

However, unless they are strictly territorial, their ranges can overlap, so unlike the human example, they can choose to live in a preferred area even though another animal is already there. As animal density increases in the most preferred habitat, however, resources become less available to each individual, so the quality of the habitat from each resident's perspective diminishes. Thus unless individuals benefit from the presence of others, their home range selection is negatively influenced by conspecific density. Other competing species have a further interacting effect on habitat availability and hence selection. Because competition changes each animal's perception of habitat availability, human measurement of availability, based on the assumption of free and equal access, is inevitably inaccurate. Animals tend to be more uniformly distributed across patchy landscapes than predicted from studies of habitat selection.

Another major problem in measuring habitat availability is the recognition and treatment of areas of nonhabitat that may exist within home ranges. Part of the difficulty arises simply because our concept of home range is too nebulous. Home range is typically defined as the area used by an animal for its normal activities, but home range area is a human perception, not a biological entity.

Humans may perceive the landscape as a mosaic of habitats that fit together like a jigsaw puzzle, on which are superimposed home ranges of animals. In contrast, animals may perceive the landscape as series of corridors or islands sprinkled in an ocean of nonhabitat. If we unwittingly define available habitat from our human perspective, and include large patches of nonhabitat that the animal does not really perceive as among its choices of places to live, a comparison of use to availability might demonstrate nothing more than avoidance of the nonhabitat.

This would be grossly accurate, but not particularly insightful. An example was presented by Johnson, where mallards (Anas platyrhynchos) rarely used open water areas far from shore, but the area of open water was large. Standard means of comparing use to availability, such as the chi-square test, might show open water to be avoided and all other habitats selected; however, a knowledgeable duck biologist would recognize this as a trivial result, and might elect to exclude this obvious nonhabitat from the analysis. Other cases may not be so clear-cut. Manly *et al.* presented an example with California quail (Callipepla californica), taken from a study by Stinnett and Klebenow. Bonferroni confidence limits, and hence perceptions of selection, depended on whether habitats that were not used as escape cover when the birds paired for mating were included or excluded from the analysis. In this case the habitats that were not used as escape cover during mating were not obvious nonhabitats because the birds used them in other circumstances and for other activities.

In this example, habitats 1, 3, and 4 were used and thus also available. Habitat 2 (depicted as a swamp) appears to have been traversed, possibly just to get from habitat 1 to habitat 3; if it was used simply because of its location, not because of its habitat-related attributes, a question arises as to whether it should be considered in the analysis. Conversely, although habitat 5 was not used, it may or may not be considered available.

Judged within the context of the home range boundaries, point A in habitat 5 appears to be unavailable, yet this point is closer to known locations of the animal than points B or C, which are both within the apparent home range. Habitat availability is a nebulous concept, and thus may be difficult to measure. Similarly, although the figure depicts a travel route, from which relative use of habitats might be deduced, most analyses deal with relative time, not distance, in each habitat (partly because telemetry data are generally comprised of point locations); it is unclear which is really a better measure of use. An advantage of Johnson's technique is that the results are rather robust to inclusion or exclusion of habitat types that are rarely used. A problem with Johnson's technique is that because it is based on rankings of use and availability, habitats will not appear to be selected if their proportionate use is ranked the same as their availability.

Thus even if the animal spends an inordinate amount of time in the habitat that is most available, selection for this habitat will not be detected using this technique because both use and availability are ranked the same. The Manly—Chesson index of habitat selection also does not fluctuate with inclusion or exclusion of seldom-used habitats, and Manly *et al.* showed that this index is much more versatile than Johnson's in many other respects. Recently, it was adapted by Arthur *et al.* to handle situations in which habitat availability changes. These authors recognized that habitats available to polar bears (Ursus maritimus) varied with changes in ice conditions and with movements of bears across their enormous home ranges.

Thus they defined availability separately for each radiolocation, using the habitat composition of a circle with a radius (from the radiolocation) equal to the expected distance a bear would travel during the time between radiolocations; habitat availability within these circles was then compared with the type of habitat the bear actually used the next time it was located. Another attribute of Manly *et al.*'s procedure is that it can be used to analyse data from site attribute studies as well as use—availability studies, although site attribute studies also face problems in assessing availability. If used sites are compared to random sites, the universe from which the random sites are drawn must be defined. As discussed earlier, that universe can be some arbitrarily defined study area, a composite home range of study animals, or each individual home range. Additional difficulties may arise if the comparison is between used and unused sites because errors may arise in distinguishing unused sites (*i.e.*, nonobservation of use may not mean nonuse).

Furthermore, unused sites may be vacant for a variety of reasons, some of which are unrelated to the physical habitat (*e.g.*, human disturbance, exploitation, predation, parasites, interspecific competition). Some predictive models have fared poorly when they did not consider such variables. Geffen *et al*. found, unexpectedly, that Blanford's foxes (Vulpes cana) in desert environments were rarely observed near springs, where water and food were most abundant, probably because this habitat was favoured by and provided cover for potential predators. In order to assess the criteria used by a species in selecting sites, investigators ideally should choose for comparison sites with both available resources and predators (or other confounding agents) present, as well as sites with only one or the other; however, such comparisons are unavailable in most field studies.

If a species is very selective in its choice of sites, differences between used and unused sites may be quite subtle; these subtleties would not be discernible in site attribute studies if the investigator chose unused or random sites that were very different from the used sites. The scale of comparison in this case would be too coarse. In an attempt to circumvent this difficulty, Capen *et al*. eliminated available sites in habitat types that were "radically different" from those that were used (analogous to eliminating nonhabitats in use—availability studies). Conversely, if in attempting to use a finer scale of comparison one picked random sites from too narrow a universe, such that they were all very similar to the used sites, habitat differences might not be detected if a large portion of the random sites were used. This points out the advantage of distinguishing unused sites instead of just random sites and of selecting unused sites that are similar in many respects to the used sites. Use—availability studies do not distinguish unused areas and so may be especially prone to problems of too fine or too coarse a scale of comparison. The coarse-scale problem (used and available areas are too dissimilar to detect the true basis for selection) may occur when composition of home ranges or habitat use within home ranges is compared to some broader study area. The fine-scale problem may occur when habitat use is compared to availability within home ranges.

Thus these two scales of comparison may yield different results. McClean *et al.* examined the effects of varying the definition of available habitat, from the entire study area to progressively smaller-sized circles around individual radiolocations. They found that selection became increasingly difficult to detect as availability was defined by a smaller and smaller area. This result is not surprising because the radiolocation represents use, so habitat composition within smaller areas around each location more closely matches areas of actual use.

ASSESSING HABITAT SELECTION

Perceived habitat selection may vary with the technique chosen to compare use and availability or to compare attributes of used and unused (or available) sites. Some of this variation in perceived selection stems from the fact that different methods actually test different biological hypotheses and some is from the different assumptions inherent in these techniques and their sensitivity to violation of these assumptions. Manly *et al.*'s technique can handle both use—availability and site attribute study designs. Moreover, it can be performed on an individual animal basis or with pooled data, it can be used to compare habitat selection among groups (*e.g.*, species, sex—age classes, seasons, times of day, times within seasons), and it can incorporate both discrete and continuous variables. For these reasons, it has been heralded as a unified approach.

Manly *et al.*'s approach generates a resource selection probability function, giving the probability of a site being used as a function of various habitat variables. Each habitat variable can be tested to determine whether it contributes significantly to the probability of use. In the special case of only a single categorical habitat variable (*i.e.*, habitat type), the function reduces to the Manly—Chesson selection index. An advantage of this index, as discussed earlier, is that it is rather unaffected by the inclusion or exclusion of seldom-used habitats. In this sense, Chesson suggested that the index is a measure of preference that "does not change with density unless behaviour changes" and that it represents the expected use of the various resources if all were equally abundant.

Consider first the simple example presented by Chesson to demonstrate the intuitiveness of the Manly—Chesson technique. The example deals with choice of foods. Suppose habitats A and B are equally available, and an animal spends 25 per cent of its time in habitat A and 75 per cent in habitat B. Because the Manly—Chesson selection index represents the expected use when resources are equally available, the index for each habitat in this case simply equals their proportional use (0.25 and 0.75 for A and B, respectively). Now suppose that the same animal is placed in an area composed of 80 per cent habitat C and 20 per cent habitat B, and it uses C 40 per cent of the time and B 60 per cent.

The Manly—Chesson index would be 0.14 for habitat C and 0.86 for habitat B, suggesting that if habitats C and B had been equally available, they would have been used in these proportions. Because both A and C were compared against the same standard (habitat B), the results indicate that A would be preferred to C if those two types were offered together. However, given that the animal used A only 25 per cent of the time but C 40 per cent of the time, when in both cases the other choice was habitat B, the higher standardized selection index for A is not intuitive; these results are clearly a function of the higher availability of habitat C. A fatal flaw of habitat selection studies in general, especially use—availability studies, is that they are based on the assumption that the more available a resource is, the more likely an animal should be to use it. This may not be true at all, may be true for only some resources, or may hold only within a narrow range of availabilities. Manly *et al.* made the explicit assumption, applicable for all models that availability remains constant for the period of study (if availability changes seasonally, data can be analysed by season). This may seem like a benign assumption, but in reality it masks a fundamental weakness of the process. Of what value are measures of selection if they are specific to a single array of habitats?

Measures of selection are supposed to be reflections of inherent preference—expected choices when availabilities of all habitat types are equal—so if selection appears to change as availability changes, then preference cannot be inferred from perceived selection when availabilities of habitats are unequal. In other words, if the goal is to assess habitat preferences for a population of animals based on habitat selection observed among a collection of individuals in that population, then something is amiss if selectivity appears to differ among these individuals simply because they have different habitat compositions available to them.

Consider a human analogy that demonstrates the effects of changes in availability on perceived selection. While at home a person spends 50 per cent of the time sleeping and 20 per cent preparing food and eating meals in the kitchen; the bedroom occupies 20 per cent of the area of the house, and the kitchen 10 per cent. Manly—Chesson selection indices for these rooms would be 0.51 and 0.40, respectively. Now suppose the person feels cramped in the kitchen and moves a wall, making it twice as big, at the expense of a room other than the bedroom. Afterwards the kitchen makes up 20 per cent of the area of the house, the same as the bedroom, but use of the kitchen does not increase because it still takes the same amount of time to prepare and consume meals there.

The selection index for the kitchen thus drops to 0.25, despite the fact that it is now more comfortable and better serves its purpose. Moreover, although no changes were made to the bedroom, its selection index improved to 0.63 as a result of the renovations to the kitchen. Superficially, it would appear that the expense for remodeling was not worth it. Analogously, one might imagine a situation in which an animal used a habitat substantially more than its availability, but used it only for sleeping. If that habitat became more available, the animal would not be expected to sleep more, so its selection for it would appear to decline. A management agency that produced more of this habitat because results of a habitat selection study showed it to be used disproportionate to its availability would be disappointed to find that these efforts made the animal's selection for it drop.

These examples demonstrate cases in which the activity requires a fixed amount of time, so increasing availability of the preferred setting for that activity has no effect on how much time is spent there. This situation is just a special case demonstrating the point that use and availability are not inexorably linked. In the example of the house, the renovated kitchen might entice the person to spend more time there, but only up to a point (one certainly would not sleep there).

Conversely, if the dining room had been remodeled at the expense of room in the kitchen, the person might not eat in the kitchen anymore, but, no matter how small it was, still prepare food there. Each room might thus have its own functional relationship between area and use. Similarly, if an animal prefers a certain habitat for resting because it offers protection from predators, it might spend more time resting in a larger patch of that habitat because it offers greater safety than a small patch. Enlarging a patch that offers virtually no predator protection to a size yielding some predator protection might thus cause significantly increased use of the patch; however, additional enlargements might have progressively lesser effects on use because they do not add much predator protection, and eventually further enlargements do nothing, or might even attract a different predator, thus deterring use.

Various scenarios and corresponding relationships between patch size and use are plausible. Considering that the relationship between patch size and use probably varies among habitat types and the mathematical relationship between use and availability also differs among the various selection indices, it seems doubtful that one could assess selection just by comparing relative use to the relative area of different habitats. Mysterud and Ims proposed a logistic regression model to compare use:availability ratios among study subjects that had differing habitat compositions available to them. This model thus provides a test of the assumption that use increases with increased habitat availability. Their method is applicable to cases in which habitats can be categorized into two discrete types (*e.g.*, forested vs. nonforested, oak vs. nonoak). They reexamined two data sets that Aebischer *et al.* had analysed using compositional analysis.

In one, use increased with increased availability of a habitat for 9 of 12 ring-necked pheasants (Phasianus colchicus); however, three individuals did not fit this trend. In the second example, gray squirrels (Sciurus carolinensis) showed an inverse relationship between use and availability of open habitats within their home ranges (the same unexpected relationship therefore existed for the alternate, forested habitat). It was surmised that size and interspersion of habitat patches greatly affected the choices that these animals made, more so than just total habitat area.

Similarly, Mysterud and Ostbye found that although roe deer (Capreolus capreolus) in winter chose open canopy habitat for feeding and dense canopy for resting, they had to balance the advantages of being in each type of habitat against the energetic disadvantages of traveling between them, so patch size (distance between patches) affected habitat selection.

Mysterud *et al.* suggested that for animals such as roe deer, which face tradeoffs in using different habitats, selection is not directly related to resource availability, so habitat rankings based simply on ratios of use to availability often are misleading. Bowyer *et al.* used a site attribute analysis to examine habitat selection related to various tradeoffs faced by black-tailed deer (Odocoileus hemionus).

Assessing selection can be extraordinarily complex because each habitat is not a single patch, but a series of patches of different sizes and shapes, each bordering other patches of different sizes, shapes, and habitat types.

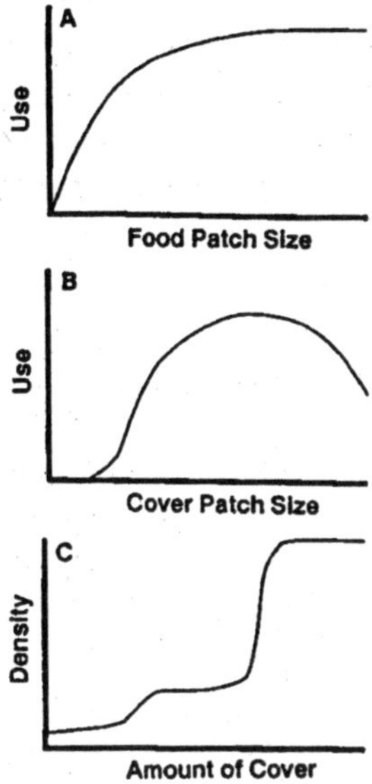

Otis offered a model that tests for the disproportionate use of habitat types as well as habitat patches, thereby providing a means of assessing things such as minimum patch size requirements. Data for this model (patch size distributions for each habitat type and locations of animals in specific patches) are available with modern geographic information system (GIS) coverages. This model still does not take into account habitat interspersion and juxtaposition, which probably have significant effects on selection for many species. For example, Porter and Church found that a standard use—availability analysis of habitat selection by wild turkeys indicated an avoidance of agricultural lands, when in reality turkeys used agricultural lands extensively, but only those near hardwood forests.

In essence, the turkeys viewed the edge between field and forest as a separate habitat type. Similarly, Neu *et al.* posited that moose might feed preferentially in a recent burn, but not too far from the surrounding forest. Thus they defined four habitat types—the interior of the burn, the burn periphery, the forest edge adjoining the burn, and the remainder of the forest—and through a simple chi-square analysis showed selection for the edge. Most situations probably are not this simple. Many authors have admitted to the importance, but difficulty, of incorporating spatial aspects of habitats in use—availability analyses. Porter and Church proposed a method whereby the study area is gridded into cells and an assortment of habitat variables within those cells are examined through multivariate analyses to find those that best explain differential use of cells. Litvaitis *et al.* did just that in a study of bobcats (Felis rufus), which predated the paper by Porter and Church. Litvaitis *et al.* looked for associations (using regression and DFA) between the number of radiolocations within 25-ha cells inside home ranges and measurements of several habitat variables sampled there; however,

They found that these habitat variables poorly explained variation in frequency of use. Servheen and Lyon used a similar approach in assessing habitat selection by caribou (Rangifer tarandus). They measured habitat variables in 40-ha circles around telemetry locations and sought to find those that best differentiated the areas that the animals used seasonally. Although they had no real measure of juxtaposition or interspersion of habitats, their 40-ha circles contained habitats neighbouring the one actually occupied, so the composition of these circles gave an indication of habitat combinations that corresponded with seasonal use. In another similar approach, Clark *et al.* used grid cells that could encompass several habitat types near the locations of radiocollared black bears.

A suite of habitat characteristics (including the number of different habitat types) within each cell used by bears were combined to form what they called an ideal habitat profile. The habitat quality of each cell in the study area was then assessed by comparing it to this hypothetical ideal cell. Each of these studies looked at differential use, rather than use in terms of availability, and thus avoided the fatal flaw of habitat selection studies. Site attribute studies are like the habitat use studies just discussed, except that instead of comparing cells with varying degrees of use, they categorize cells (sites) simply as used or unused; based on this, important habitat variables are identified. Interspersion and juxtaposition of habitats can thus be investigated. For example, Coker and Capen examined cowbird (Molothrus ater) selection for habitat patches of various size, shape, and location relative to other habitats by entering these variables in a logistic regression with use (used or not used) as the dependent variable.

Similarly, Chapin *et al.* compared habitat variables (including an index of the extent of habitat edge) in grid cells of different sizes that were used (*i.e.*, had at least one telemetry location) by American martens with those in cells not used by martens, and also compared characteristics of forest patches that were used and not used. McLellan argued that observed use is a better indicator of habitat selection than use relative to availability. He reasoned that an animal familiar with its home range knows the availability and location of resources, so an animal's location at any given moment represents selection.

He gave an example of a person at a buffet selecting a slice of beef from a 500-kg steer and an equalsized slice of pork from a 100-kg pig; based on use alone, pork and beef were selected equally, but compared to availability, pork appears to be selected over beef, which is obviously absurd. However, had the steer and pig been cut up in equal-sized chunks and distributed over a large area, and after considerable searching the person still returned with an equal quantity of the two foods, active selection for pork would indeed seem apparent.

The key difference is that in the latter case the person had to search for the food; selection was evidenced by the extra effort expended in finding the pork (and apparently bypassing chunks of beef). This searching for resources is really the basis for the development of use—availability comparisons and explains why it originated with studies of diet. In most cases animals do not know the location of all foods in their home range, so dietary selection based on availability may be appropriate. However, habitats are not spread around like chunks of pork and beef, but occur in large patches, the locations of which are known by the animals; thus habitats are probably more like McLellan's whole steer and whole pig than the cut up chunks of meat spread randomly around.

Consider some actual examples of how observed use and use versus availability can lead to disparate interpretations of selection. Prayurasiddhi investigated use and selection among two large ungulates, gaur (Bos gaurus) and banteng (B. javanicus), in Thailand. He differentiated two general study area boundaries, one of which more closely matched the area that his radiocollared animals used most intensively. He also used actual home range boundaries as a third representation of the study area and hence the available habitat. He found that this variation in the area considered to be available habitat resulted in drastic differences in perceived habitat selection.

One habitat that received 46 per cent of use by gaur was deemed to be selected for, whereas another habitat that received 45 per cent of use was seemingly "selected against." In one case banteng were judged to be unselective in their use of a habitat in which they spent 75 per cent of their time. Prayurasiddhi recognized these difficulties and decided to evaluate seasonal changes in habitat selection and habitat-related differences among species based on use alone, rather than use compared to availability.

In another analogous situation, Macdonald and Courtenay found that crab-eating zorros (foxes, Cerdocyon thous) in Amazonian Brazil spent most of their time (64 per cent) in wooded savanna and scrub habitats; however, because these two habitats were abundant within the home ranges of the study animals, they ranked the lowest in apparent preference (sixth and seventh among seven defined types) based on a comparison of use to availability. During the wet season, however, when lowland habitats flooded, the zorros used upland wooded savannas and scrub habitats even more, making the apparent preference. Ranking for these rise; in reality, the area of available (not flooded) lowland habitats diminished, but this reduction could not be measured and therefore was not taken into account in the use:availability calculations. The authors realised that the seasonal difference in apparent habitat preferences was thus an artifact of unmeasured changes in availability.

In another such case, Garrett *et al.* found that tidal flats were the principal foraging habitat for bald eagles (Haliaeetus leucocephalus) and observed that nearly one-fourth of their perch sites were within this habitat. That is, based on use alone, this area was clearly attractive to these birds. However, because this habitat was so widely available, especially at low tide, a comparison of use to availability suggested that the eagles avoided it. Apparent preference thus changed radically with tidal fluctuations.

A problem with assessing selection from use alone is that use often does change with availability. In the eagle study, tidal flats were rarely used during high tide because they were largely unavailable. The eagles may have preferred tidal flats to other habitats, but use of these areas depended on availability, up to a point. When availability exceeded some threshold they did not continue to increase their use, and when it declined below some threshold their use of the area ceased. Between these thresholds some relationship may exist between use and availability. However, relationships between use and availability probably vary among habitat types, and within each habitat may depend on what other habitats are available.

It might seem that site attribute studies avert these problems because selection is inferred not from differences between use and availability, but from differences between used sites and other sites. However, the comparative sites are taken proportionately from those available, so the comparison is still, in a sense, the same as in the use—availability design.

Consider a study that locates 10 nest sites in each of two available habitat types and then compares the characteristics of these to 20 randomly chosen points. If one habitat type is much more common than the other, then the 20 randomly chosen points fall mostly within the more common habitat; this sampling could result in possible differences between the average characteristics of the randomly chosen sites and the real nests, suggesting that the species selected for nest sites in the less common habitat type. This logic falters, though, for the reason highlighted in McLellan's example of the buffet, in which a person chose equal portions of beef and pork. In that example, the equal choice of the two types of food was clear evidence of nonselection, or equal preference, regardless of availability. Accordingly, in the hypothetical case of the 20 nests, one should question why, if there was strong selection for one habitat type, did an equal number nest in each type? The conclusion that this represents evidence of preference for the less common type is (just as in use—availability comparisons) based on the assumption that if individuals had no preference, they would chose sites.

Randomly, which is typically thought to mean in proportion to availability. However, suppose each individual, having no preference, just flipped a coin, so to speak, to decide which of the two types of habitats to nest in each year. The result would be as posed in the example, an equal number nesting in each type. There are two other problems (really two aspects of the same problem) with detecting selection using a site attribute design. James and McCulloch showed that a species could be highly selective for a certain habitat component, as indicated by a low variance among values of this component at the sites it chose, but if the mean value for the component at selected sites approximately equaled the mean in the available habitat (where the variance was much higher than at selected sites), then selection might not be detected (type II error).

Conversely, Rexstad *et al.* tested multivariate statistical procedures that are commonly used in site attribute studies and found that significant results, which in a real study would be indicative of selection for certain habitat components, were often derived from a collection of meaningless data (type I error). Taylor attributed Rexstad *et al.*'s spurious. Results to ill-defined hypotheses and inappropriate use of statistical procedures; however, Rexstad *et al.* showed that their study mimicked the majority of published applications of these techniques. North and Reynolds expanded on Rexstad *et al.*'s concerns that the statistical procedures used in site attribute studies may yield misleading results because assumptions are commonly violated.

INFERRING HABITAT QUALITY

If habitat selection studies were simply an attempt to better understand behaviour and natural history, errors would merely be a setback in scientific inquiry. However, these studies are typically designed to provide management guidelines. Habitats are managed based on supposed importance in terms of fitness for members of the target population. High-quality habitats, by definition, produce animals with high reproduction and survival, and hence a population with a high growth rate (and high yield, for harvested species).

However, the assumption that we can infer habitat quality or suitability from studies of habitat selection—that selection, even if accurately measured, is directly related to each habitat's potential contribution to individual fitness and hence the population's growth rate—represents what the second fatal flaw in the process of habitat evaluation. One of the best examples of a purported relationship between habitat selection and fitness was provided by a study of leaf-galling aphids (Pemphigus betae) that parasitize narrowleaf cottonwood trees (Populus angustifolia). These aphids overwinter in tree bark, then over a short period in the spring migrate up the tree and become entombed in expanding leaf tissue, where they reproduce.

Whitham found that aphids apparently selected large leaves over small leaves, even though large leaves were less available (less than 2 per cent of leaves were over 15 cm; more than 30 per cent were 5 cm or less); moreover, those that colonized large leaves had better survival and reproduction than those colonizing small leaves. From this evidence it appeared that aphids had the ability to select leaves that offered the highest fitness; if not, the size distribution of colonized leaves would have matched the size distribution of leaves available on the tree (*i.e.*, use would have equaled availability). However, Rhomberg observed that large leaves tended to be those near the tips of twigs (where they receive more light), and these opened a few days later than basal leaves. When the aphids begin their migration, basal leaves are already too mature to enable them to form a gall, so they must colonize the more distal leaves; thus, the basal leaves are in essence nonhabitat. The migration appears to be timed so that the aphids inevitably colonize leaves that are destined to be favourable to their fitness, but individually, they do not select these among less favourable habitats.

This is not to say that other animals do not select habitats favourable to their fitness; certainly they do. But if the comprehensive and seemingly compelling data on a simple system such as aphids on cottonwood trees can be misinterpreted, then clearly there is much room for misinterpreting relationships between habitat selection and fitness in more complicated systems. The presumed link between selection and fitness is rarely tested, partly because such tests are difficult, but also because this relationship is viewed more often as fact than as assumption. However, there are many reasons why this relationship may not hold. Habitats that are used infrequently may be more important than suggested by the time spent there.

Conversely, habitats used preferentially for activities that require a lot of time (*e.g.*, resting) may be less important than indicated by their use; a resting habitat may be substitutable, with little effect on fitness, whereas a specific foraging habitat (or a water hole in an arid environment) may be more essential, even if not used very much. Differentiating activity-specific habitat use may help alleviate this difficulty, although this, too, is seldom done, and moreover, does not necessarily circumvent the problem. For example, Powell studied winter foraging of fishers, which preyed heavily on porcupines (Erethizon dorsatum) but spent a disproportionately small amount of time hunting in upland hardwood habitats where porcupines were common. The reason, he found, was that fishers rapidly located porcupines at known den sites, thus minimizing their search and chase times. In contrast, fishers had a harder time hunting snowshoe hares, so they spent a larger amount of time in lowland conifer habitats where hares were more common.

Clearly, the relative importance of the two habitats to fishers was not reflected by their time spent hunting in each. For animals that feed on a variety of different foods, a mixture of different habitats may be more beneficial in many respects than a single, highly preferred type, and the time spent in each may be a poor indicator of importance of either the specific type or the overall mix. Omnivorous Egyptian mongooses (Herpestes ichneumon), for example, favour one specific habitat for resting, but use a mix of habitats when feeding.

Fitness also may be affected by predation and interspecific or intraspecific competition. These factors can change the cost:benefit ratio of a habitat and hence alter an animal's habitat use. Cowlishaw found that baboons (Papio cynocephalus) spent more time feeding in a habitat with poor food but a low risk of predation than in a food-rich habitat with a high risk of predation.

Fitness was probably enhanced by this choice of habitats. Fitness certainly was enhanced for a herd of caribou in Ontario, Canada, that spent most of the time on an island. The nearby mainland had higherquality forage but also a high density of wolves. Ferguson *et al.* found that the island occupants sacrificed nutrition, which was reflected in smaller body and antler size as well as starvation of their calves, for overall higher survival, which enabled this population to persist while herds that remained on the mainland perished from predation. One could argue that the animals on the mainland made poor choices in habitat selection. Animals may not always correctly perceive risks, especially in a novel or changing situation, so their choice of habitats may not necessarily be best in terms of their fitness.

Pollution is one such risk that is rarely evaluated in habitat studies. Consider Mallory *et al.*'s study of the effects of acid rain on habitat quality for common goldeneyes (Bucephala clangula). Acid rain killed fish in lakes, which enabled proliferation of invertebrates, thus providing more food and enhanced reproduction for goldeneyes. However, these ducks may inevitably suffer other consequences of the acidified habitat. In some situations the choices made by animals may maximize fitness over the long term (*e.g.*, if the changes are ephemeral); in other cases animals may require time to adapt. Suppose that an animal uses alternate habitats more often when the normally preferred habitat becomes less profitable because of increased density and hence competition for resources. If the alternative habitats are more available than the preferred habitat, then use of these alternative habitats can end up higher than that of the otherwise preferred habitat. Situations like this seemingly justify considering habitat use in terms of availability.

However, Hobbs and Hanley showed both intuitively and with computer modeling that intraspecific competition may alter use: availability ratios such that they may not reflect the true quality of a habitat in terms of its potential contribution to population growth. Certainly interspecific competition and predation are integral components of habitat quality, and managers concerned about a given species should not evaluate habitats independent of these factors.

However, perceptions of habitat quality that are affected by density of the targeted species are an artifact of the method. Recall, as Peek noted, that habitat preference should be innate, and intricately tied to fitness; if our perception of preference declines as the preferred habitat becomes more densely populated, then our perception is clearly mistaken. Hobbs and Hanley concluded that "use/availability data inherently reflect differences in animal density among habitats" (independent of social behaviour) and therefore reveal little about habitat quality.

Consequently, they recommended abandonment of the use—availability design in favour of an approach that enables a direct investigation of the link between habitats and fitness. Site attribute studies often focus on sites of biological importance (as reviewed earlier) and therefore may provide more direct insights into habitat variables that affect fitness. However, as pointed out by North and Reynolds, conclusions from studies of this sort are based on two fundamental, but questionable, underlying assumptions: that used sites are in suitable habitat and that unused sites are in unsuitable habitat.

Consider the study by Meyer *et al.*, who observed significant habitat-related differences among sites that were occupied versus unoccupied by spotted owls (Strix occidentalis).

They found, as expected, that these owls tended to select sites having a large proportion and large patches of old-growth forest, but suggested that the owls could tolerate certain amounts of clearcut and young forest (based on the fact that used sites contained such areas). Their reproductive data were inconclusive about what the owls could tolerate from a population standpoint, but Meyer *et al* nonetheless asserted "that resource selection probability functions are a reliable tool for assessing capability of landscapes to support northern spotted owls." Hence they concluded that their habitat selection model "can be used to predict the probability that a given landscape mosaic will be a suitable spotted owl site" (emphasis mine).

To avoid assumptions related to suitability in site attribute studies, North and Reynolds proposed that instead of comparing habitat characteristics of used sites with unused or random sites, used sites be categorized by their intensity of use (*e.g.*, 1–2 per cent of locations = low use, 3–10 per cent medium use, more than 10 per cent = high use), and habitat characteristics compared among such categories (*i.e.*, excluding unused sites) with a form of logistic regression. Thus this is not really a site attribute design (according to my definition), nor is it even applicable to the types of situations in which site attribute studies are commonly used (*i.e.*, in which an animal remains at a single site for a fairly long period of time).

Rather, this procedure represents just another means of investigating habitat characteristics associated with differing degrees of use. Whereas these methods may avoid many of the statistical and general assumptions of site attribute studies, they digress further from the issue of habitat-related fitness because the studied sites, instead of being of some special biological significance, are simply points where the animal was located some number of times. As North and Reynolds recognized, intensity of use may not be a good indicator of habitat quality.

ADVANTAGES AND PROBLEMS OF THE DEMOGRAPHIC RESPONSE DESIGN

The best measure of habitat quality would be a test of its effects on demographic parameters such as population growth and carrying capacity. Such tests are extraordinarily difficult in most situations, as evidenced by the scant published studies of this nature. Because habitat-specific density is actually a reflection of differential habitat use, investigations of habitat-related density suffer the same drawbacks as studies of habitat use. Density tends to be an ineffective measure of habitat quality because it may fluctuate widely, is subject to sizable errors in estimation, and may be largely influenced by social factors. Van Horne gave several examples of situations in which juveniles were restricted from settling in the best habitats and thus accumulated in large numbers in poorer quality habitats.

She indicated that such circumstances are likely to be common among generalist species with high reproductive rates and a social hierarchy. For these species in particular, then, habitat-specific density would probably be a poor indicator of habitat quality unless the population is well below carrying capacity. A good example was provided by Messier *et al.*, who showed that density of muskrats (Ondatra zibethicus) during a general population increase swelled 30- to 90fold in low-quality habitats but much less in high-quality habitats. Considering that, in general, animals in poor-quality habitats should be trying to leave and ones in high-quality habitats trying to stay (and keep competitors out), Winker *et al.* posited that turnover rate would be a better index of habitat quality than density. They measured turnover rates for wood thrushes (Catharus mustelinus) by examining recapture rates and telemetry movement data; low-quality habitat was defined as that in which recapture rates were low and many radiotagged birds were transient visitors.

They found density and habitat quality, assessed in terms of turnover rates, to be inversely related. Notably, Winker *et al.*'s turnover rate model may not be applicable to other species, even other territorial, noncolonial songbirds, some of which are preferentially attracted to habitats occupied by conspecifics, which they use as a cue to habitat quality. Conspecific attraction tends to perpetuate use of the same areas across generations, so even if habitat quality deteriorates, high densities may be maintained through tradition. Other competing species or unidentified confounding variables also may weaken the linkage between habitat quality and density. Maurer measured density and various habitat characteristics for five species of grassland birds; the habitat models developed to explain species-specific density in one study area were inexplicably poor predictors of density in a nearby area with similar habitat.

Kellner *et al.* found that density was positively related to reproductive success in only 7 of 17 bird studies that they reviewed; more studies showed a negative relationship. Sherry and Holmes felt that density should be relied on as an indicator of habitat quality only if it is corroborated by other data, as was the case in their study, in which population density and weight loss of wintering migrant birds were correlated (high weight loss in areas with low densities) and both were related to habitat type. Reproduction and survival data may be more apt to reflect real influences of habitat on demographics. However, reproduction and survival are probably also tied to habitat in a complex manner. For example, a number of studies observed a direct relationship between cover and the survival (and thus density) of voles (Microtus spp.), but a lower threshold exists below which reductions in cover have little effect on vole density; above the threshold, survival and density increase but eventually reach an upper asymptote.

The vole studies found that cover provides food as well as protection from predators, and also may affect microclimate, activity patterns, and interactions among conspecifics, all of which affect the cover—density relationship. Moreover, male and female voles have different responses to varying cover, and cover—demographic relationships tend to be different for other small grassland rodents.

Each of the various habitat components that relate to an animal's fitness probably has thresholds, asymptotes, and inflection points, and these limits may vary with the mix, shape, size, and juxtaposition of habitat components available; however, few attempts have been made to assess any of these factors individually, *let al*one in combination. Several studies also found that density-dependent effects may reduce reproduction or survival independent of habitat quality or may even result in higher fitness in low-quality, less crowded habitats. Zimmerman found that nesting success of dickcissels (Spiza americana) did not differ between habitats and was unrelated to density, but females nested preferentially in the habitat preferred by males; males chose this more heterogeneous habitat because they could sequester more nest sites and thus mate with more females.

In other studies, reproduction and survival were found to be unrelated to measured habitat variables, despite evidence of habitat selection, possibly because of confounding effects of weather, human disturbance, measurement error, and other factors. Even experimental demographic response studies have been plagued with unexpected variations in confounding variables. Nevertheless, some carefully designed studies observed links between habitat and reproduction or survival and through further investigation discovered the underlying causes.

A final major problem, discussed previously in relation to use—availability and site attribute designs, is scale. Levin showed clearly that there is no single correct scale for studying ecological relationships. Animals view and react to their environment at various scales. Human perceptions of ecological systems are inescapably biased or incomplete because they are filtered by the observational scale chosen for the investigation. Some demographic response studies have recognized this and have adopted a multiscaled approach. For example, Orians and Wittenberger found that densities of yellowheaded blackbirds (Xanthocephalus xanthocephalus) were higher on marshes with higher food (insect) abundance but that density of blackbird territories within these marshes was related to vegetational structure, not food.

The authors postulated that these birds may not select food-rich territories because they often hunt outside their territories. Also, they establish their territories before the full emergence of insects and hence may not be able to predict future food abundance on a scale smaller than the marsh level. Pedlar *et al.* developed habitat models on two different scales to explain variation in raccoon (Procyon lotor) density. One model was fit to macrohabitat features and the other to microhabitat variables, after which the two were combined to form a more comprehensive model relating density to habitat at both scales.

Morris observed both macrohabitat and microhabitat differences among several species of small mammals, suggesting habitat selection on both scales, but found that variation in density was much more evident at the macrohabitat level. On a larger scale, Dooley and Bowers discovered, counter to their expectations, that densities and population growth rates of voles were higher in patches within a fragmented landscape than in an unfragmented landscape, whereas total population size was higher in the unfragmented landscape (because more total habitat was available).

Landscape fragmentation caused overall habitat loss but, on a finer scale, enhanced reproduction within individual habitat fragments. Similarly, Brown and Litvaitis noted that some mammalian predators that hunt preferentially in forests nonetheless often exist at lower densities in homogeneous forests than in forests interspersed with disturbed areas. Each of these examples represents cases in which a single-scale investigation would have failed to detect habitat variables affecting population demography.

The previous examples all concerned spatial scale. Time scale may be equally important. The demographic value of a habitat may become evident only in the long term, after a population has been subjected to the stresses of a periodic drought, severe winter, or failed food crop.

APPLICATIONS AND RECOMMENDATIONS

A great deal of effort continues to be invested in habitat-related studies of wildlife. In the United States, federal land management agencies in particular have focused on developing formalized procedures for evaluating habitat for wildlife. The procedures that have been adopted rely on models derived by species experts, who in constructing these models tend to rely more on experience than on empirical data. Therefore, the models are really hypotheses in need of testing. However, because these models hypothesize explicit relationships between habitat attributes and animal populations (so-called habitat suitability indices), they cannot be rejected or accepted in normal scientific fashion; that is, none of the relationships are likely to be exactly correct. Thus it seems inappropriate to suggest, as is common parlance, that they should (or even could) be "verified" or "validated."

Several attempts have been made to test such models. Often this process is circular, involving just another panel of experts making qualitative assessments. In other cases, models have been tested using results of a study on habitat use or use relative to availability, with the inevitable associated shortcomings.

In some instances, habitat management prescriptions based on "common knowledge" or expert opinion have, through collection of better data, been proven faulty. One case in which model-derived habitat scores for individual home ranges were compared with reproduction, juvenile growth rates, and home range size, but no significant relationships were observed. Often, models have been tested by comparing habitat-specific densities to model predictions.

However, even if a model explains a significant portion of the variation in density, the data collected to test (or purportedly validate) the model could be better used to modify it or build new one. In most cases, habitat models have proved to be poor predictors of animal density, indicating either defectiveness of the model, lack of a clear habitat—density relationship, or effects of other confounding factors, such as hunting pressure, which are also habitat-related. Bender *et al.* found that if the variance around the estimated values of the model inputs were taken into account (a process that is not commonly done), suitability scores for a variety of habitats that appeared very disparate were not significantly different; that is, the parameter estimates were not precise enough to even enable the model to be tested.

A major inherent but generally unstated (maybe unrecognized) assumption of habitat suitability models is that high-quality habitats (*i.e.*, habitats that confer high fitness) are in fact suitable. Explicit tests of this assumption are rarely conducted, yet counterexamples exist. In one case, a habitat suitability model for Florida scrub-jays (Aphelocoma coerulescens) correlated well with demographic performance, but most of the area was found to be a population sink where mortality exceeded reproduction. Apparently, as birds competed to occupy the best habitats, they were less alert to predators, and thus suffered high mortality. In another example,

Kirsch found that interior least terns (Sterna antillarum, least terns nesting in noncoastal areas) selected high-quality nesting habitats, but possibly because of disturbance, their productivity was not sufficient to maintain population size (*i.e.*, the nesting habitats were unsuitable). Lomolino and Channell observed that remnant populations of endangered mammals often occur near the periphery of their former ranges; because the periphery of the range represents the edge of suitable habitat, studies of habitat suitability of endangered mammals based on habitat use in existing populations are likely to be misleading. These are not gratifying results.

It has not been for lack of effort, expense, or analytical developments that relationships between habitats and population growth often elude detection. It is simply the complexity of the interactions between animals and their environment that make such relationships deceptively difficult to understand.

A case in point is the spotted owl, which has undergone intense scrutiny because of its threatened status and apparent proclivity for old-growth forest in a region where the economy is tied largely to timber; despite a plethora of habitat studies, significant debate persists among ecologists as to the critical habitat requirements of this species, and why it prefers old-growth forest. However, substantive flaws in the most commonly used techniques for studying wildlife—habitat relationships apparently have not been widely recognized. It is time to reconsider the ways these techniques are used in evaluating habitat quality.

Studies of the use of habitat have merit, but also many limitations. Habitats in which an animal spends a large proportion of its time are clearly selected among others available. Even if widely available, frequently used habitats are certainly not "selected against" or "avoided" (except maybe among life forms that have no memory of where they were).

However, frequent use suggests nothing about habitat importance or substitutability in terms of fitness. Correspondingly, infrequent use may not be indicative of lack of suitability. A habitat may be used infrequently because it serves little value, because its value can be extracted in a short amount of time, because it is not readily available, or because access is constrained by threats (social pressures, competition, predation) or physical barriers. A high use:availability ratio might suggest (correctly) that such a habitat is more important than indicated by its infrequent use.

Studies of habitat use would benefit greatly from replication. Significant insights might be gained from comparisons of habitat use and use:availability among individuals, among groups of individuals in different portions of a study area, among study areas,. among time periods, and so on. If individuals are disparate in their use or apparent selection of habitats, inferences regarding habitat quality become more equivocal. In contrast, if the data are partitioned and the subsets show consistent patterns (*e.g.*, use:availability ratios for each habitat are similar among individuals despite large differences in availability within individual home ranges) or if curvilinear relationships between use and availability can be ascertained, inferences are strengthened; even so, studies of this sort provide only a superficial understanding of the effects of habitat on population dynamics. Certainly no strong prescriptions for habitat manipulation are warranted from interpretations of selection based solely on observed patterns of habitat use. Site attribute studies tend to provide stronger inferences about habitat selection. However, in identifying myriad habitat characteristics that are apparently preferred, even for activities that impinge on the animal's survival and reproduction, these studies still cannot assume that population growth would be significantly higher in more "ideal" settings. Controlled experiments can help sort out the factors important in the animal's selection processes and can thus provide a better understanding of its behaviour, but without corresponding demographic measurements, the importance of various habitat components in terms of their contribution to the animal's fitness cannot be appraised. Kirsch: "Unfortunately, proximate habitat features may not indicate habitat suitability, nor do they reveal the possible selective pressures that influence habitat selection in a system.

One must measure components of fitness, determine factors that influence fitness, and relate fitness and factors influencing fitness to habitats or habitat features." Demographic response studies are the only means of truly evaluating the relative importance and suitability of habitats for supporting animal populations. The figure depicts such a plot for 20 hypothetical home ranges (each point represents habitat use and availability in one home range) and two (of several) types of habitat (sums for both use and availability for all habitat types for each individual would total 100%).

The data suggest that increased availability of habitat 1 prompted a corresponding increase in use. In habitat 2, however, the animals showed an asymptotic relationship between use and availability; 10% availability seemed to represent an approximate threshold, above which use no longer increased. That is, if at least 10% of their home range was composed of habitat 2, the animals could obtain whatever resources they needed from this habitat by spending 10–20% of their time there. A similar threshold may exist for habitat 1 at availabilities > 50%; assumptions should not be made beyond the data.

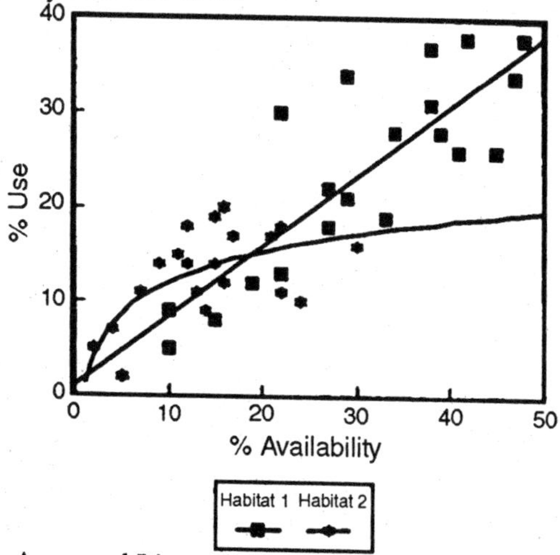

Fig. The Assumed Linear Relationship Between use and Availability of Habitats can be Tested, to an Extent, by Plotting use and Area of each Habitat in Individual Home Ranges.

These should be given greater emphasis. Oddly, Hansson suggested that habitat-specific survival might confuse perceptions of habitat selection. In his small mammal study, perceived habitat selection appeared to be a consequence rather than a cause of differential survival. It seems to me that such knowledge of habitat-specific survival is exactly the desired objective; habitat selection studies are just an indirect approach towards this end.

For many species, habitat-specific densities may be easier to measure than habitat-specific reproduction or survival, but density studies may yield uncertain or misleading results because density is the end result of various processes, both demographic and behavioural, each with potentially different habitat-specific responses. Controlled experiments should be used more often in assessing effects of habitat on demographic parameters such as density or reproduction; however, even elegant experiments may produce unexpected results in complex natural systems. Where experimentation is unfeasible, comparisons of reproduction and survival can be made among individuals or groups of individuals with different patterns of habitat use. Partridge warned that such comparisons may be confounded by competition among individuals:

Those living in preferred habitats may be dominant individuals that are naturally fitter. That is, they live where they do because they are fitter; they are not fitter because of where they live. This is an important consideration, but it is not applicable to all situations; it depends on the social structure and population size relative to carrying capacity. One way to circumvent the problem is to compare population parameters across multiple study sites with differing habitat compositions. Often, though, it may be appropriate to assume that individual physical differences are not the cause of differences in habitat use, or even if they are, that the most fit individuals are choosing the best habitats, thereby justifying comparisons within a single study site.

For studies in a single study site, an effective design would be to monitor habitat use of various individuals (*e.g.*, using radiotelemetry), and then compare their frequency of use of different habitats or the habitat composition of their home ranges to their eventual reproduction and survival. Many studies of birds have used this sort of approach to examine relationships between habitat characteristics of nesting sites and nesting success. However, few attempts have been made to relate reproduction or survival to variation in habitat use within home ranges. In one example, Aanes and Andersen observed a relationship between survival of roe deer (Capreolus capreolus) fawns and the habitat types that they used. Similarly, Aebischer *et al.* found that habitat use by individual pheasants was related to their survival.

In another example, Fuller observed that wolf (Canis lupus) territories were limited to areas with low road density (a habitat component that relates to human access) and then looked for (but did not find) a relationship between survival of radiocollared wolves and road density within individual territories.

In another study of the same species, Massolo and Meriggi found relationships between reproduction (documented presence of pups) and various natural and anthropic habitat features, including road density. In studies in which data on survival and reproduction are not obtainable or too scant for making comparisons with habitat use, surrogates (*e.g.*, body size, home range size) may, in some cases, be appropriate and could add measurably to assessments of habitat quality.

9

Evolution of Mammalian Sociality

The diverse relationships of individuals in a social network interact to create complex emergent patterns. These patterns, like the vortex that appears in an emptying sink, is not contained in the structure of a single component. Because a society represents a whole with properties different from those of its component parts, the ultimate consequences of social interactions may be remote from an observed action.

This is a fact that evolution by natural selection can take in its stride but that we, as primarily linear cause-and-effect thinkers, may find hard to accommodate. It may be clear that a lion killing a zebra is behaving adaptively, but less clear whether it is adaptive when the same lion prevents a conspecific from feeding at the kill. The immediate effect is that the first lion may have more food to eat, but the ultimate effects reverberate through a stochastically unpredictable system of long-term consequences among the whole pride.

Denied food or coalitionary aid by an ally of the snubbed individual at a later date, the fitness consequences for the originally possessive lion may be far from advantageous. An understanding of social dynamics offers insight into the adaptation of individual responses evolved from selection operating on them from the level of emergent systems.

CONSERVATION APPLICATIONS

Many problems in wildlife conservation and management involve humans causing changes to animal populations or their environment. In applied work, an understanding of the dynamics of a social system is a prerequisite to predicting the effect of human activities on, for example, spatial organization, population dynamics, and dispersal. For example, attempts to control the transmission of bovine tuberculosis by killing badgers, a reservoir of the disease, clearly disrupts the society of survivors. The effects of such perturbation on social dynamics may alter the transmission of the disease, plausibly for the worse. A similar case may be argued regarding rabies control.

Translocation of elephants without regard for the social structure that provides adolescent discipline has led to problem animals in some African parks. Tuyttens and Macdonald (in press) review some consequences of behavioural disruption for wildlife management. Population control has been shown to affect the rate and pattern of dispersal, home range size, territoriality, mating system, and the nature of social interactions in a variety of species.

UNDERSTANDING OURSELVES

So much is similar in the basic biology of vertebrates, and so universal are the processes of evolution, that an understanding of nonhuman sociality is likely to illuminate human society. This point was hitherto neglected, but stressed by Tinbergen in the foreword to Kruuk's book, in which he concluded, "It is therefore imperative for the healthy development of human biology that studies of primates be supplemented by work on animal species that have evolved adaptations to the same way of life as ancestral man."

Following Wilson's Sociobiology, it has been widely and sometimes controversially discussed. Clearly, two routes come to mind as fruitful sources of this insight: looking at the societies of species most similar to our current condition (including some hypersocial aspects that put us in circumstances to which we have not yet had time to evolve) and focusing on those currently entering evolutionary phases through which we have already passed. The first approach has prompted (or at least its promise has funded) much primatological research. The closeness of this parallel might be diluted if, as Hinde suggested, the societies of humans differ from those of other animals in that social structure in nonhumans is determined primarily by the sum of the interactions of its component individuals, whereas in human groups a structure is more often imposed from above by government or tradition in the form of Dawkins's memes.

Hinde's dichotomy may imply that the imposition of structure can cause stresses in human social systems when natural roles conflict with assigned roles. On the other hand, one could take the view that the dichotomy is not profound because the constraints of governmental ideology are loosely parallel to those imposed on all species by ecological factors such as resource dispersion. If so, a different understanding of social responses to the imposition of external constraints might be revealed by species more recently launched onto a trajectory of sociability. Certainly, badger groups show weight reduction, higher incidences of wounding, and lower reproductive success per breeding individual as group size increases. For badgers, group living may be a social innovation facilitated by the development of agriculture; individuals may be evolving towards capitalizing on this newly imposed structure (by manipulation, support, interdependence of roles, and other factors), but for the moment the stress is showing.

HOW TO DESCRIBE SOCIAL DYNAMICS

It does not detract from the excitement of behavioural, ecological, and sociobiological insights to note that recent enthusiasm for these topics has been characterized by a plethora of short, snappy papers with a clear adaptive punchline and a concomitant neglect of the empirical foundations of ethology. Historically, this arises because behavioural ecology and sociobiology were pioneered to offer ultimate functional explanations, whereas ethology embraced adaptive significance and evolution along with mechanisms and ontogeny. This vogue has led to the widespread abandonment of the ethological aspirations of the late 1970s, epitomized by Hinde's careful use of terminology and hierarchical classification to ensure compatibility between studies used for comparative work.

At its purest, this traditional ethological approach placed greater emphasis on the facility of later reinterpretation of results than on the quest for a desired result. In contrast, there is an invasive tendency to treat hard data and description as disposable assets sacrificed to analytical elegance and discussion. In a science in its infancy (such as social biology) this brings the risk that future research may be doomed to repeat previous field work solely to attempt reinterpretation of undisputed results.

ACTION, INTERACTION, AND RELATIONSHIPS

Adapting Hinde's classification, the basic units of social exchange between individual primates are action and interaction. Actions are directed towards the environment, and are often important in understanding a previous or subsequent interaction with individuals. Interactions (A attacks B) are hard to interpret without records of actions (A accidentally drops fruit, B picks up fruit). Above action and interaction in the hierarchy of social dynamics are social relationships. Relationships are quantified by the rates, frequencies, and patterning of component interactions and may be described in terms of the diversity of interactions, the degree of reciprocity or complementarity, relative frequency and pattern of interactions, synchronicity, and multidimensional qualities.

In principle, relationships may be stationary or transitionary. The former do not change with prior experience (intrinsic development) or conditions (extrinsic modification) and are at most rare and perhaps nonexistent in mammalian societies. Generalizations about relationships can be sought in various ways. Dyads may be assigned to predefined categories such as age, sex, kinship, or even personality. Personality, in this context, is a consistent moderator of interactions; for example, a shy animal tends to act differently from a bold animal under the same circumstances. Block model methods allow subgroupings to be isolated from sociomatrices.

For example, Iacobucci compared 13 methods for recovering subgroup structure from dyadic interaction data. In general, block models are based on structural equivalence of sociomatrices. Some individuals behave similarly in respect to their age, status, or sex. Using block models, these relationships between individuals can be extracted and further studied, for example, using tests for reciprocity and interchange of behaviours or using more detailed methods based on time structure of the processes.

In their study of capybara mating systems, for example, Herrera and Macdonald disentangled the effects of dominance on mating success. In that example, dominant males secured more matings than any other individual, but fewer matings than subordinate males as a class; this arose because while the dominant was busy driving off one subordinate, another sought quickly to mate with the female. The sum of social relationships may be compiled in a matrix of dyadic interactions to produce a social network. Analysis of sociomatrices assumes stationarity, which, as we have noted, is effectively nonexistent. The solution is to divide sociomatrices into appropriately defined periods that approximate stationarity. This might involve consideration of, for example, "the first 50 interactions," "the second 50 interactions," and so forth, or "wet season interaction" versus "dry season interaction," or "simultaneous presence of dominant" versus "absence of dominant." Nested analysis is a common way to achieve this goal.

SOCIAL STRUCTURE, FROM SURFACE TO DEEP

A social network provides a snapshot of one facet of society. By analogy, one analysis of a social network is akin to the view through one window into a large and labyrinthine house; the view through all windows gives the structure. It is therefore necessary to compile several networks that describe different facets of one society. This task may be made harder because different sociometric variables (*e.g.*, grooming, aggression, and play) may not follow similar patterns of stationarity; aggression may covary with age and presence of dominant, whereas grooming may not. Notwithstanding these complexities, a society's structure can be described in terms of these networks. Indeed, there are layers of completeness to this description. The structure that prevails may vary on a circadian basis, or seasonally or annually; it may also be characteristic of a species' society in only one habitat or set of environmental conditions. At its most fundamental, elements of social structure may characterize all populations of a species. Therefore, social structure might usefully be categorized on a continuum from surface structures to deep structures. The study of social dynamics seeks to describe and explain the patterning of transitions and stability of social structures. An important goal of evolutionary biology is to identify the rules, derived from a variety of empirical and theoretical sources, that are thought to guide an individual's decisions in a social context—Axelrod's seminal question of whether to cooperate. An accurate description of structure is clearly a prerequisite to a sensible exploration of these rules.

Operationally, the point at which a thorough description of social structure is complete is probably the first point at which it is legitimate to consider exchanging data language (grooms, fights) for theory language (alliance formation, competition). These interpretative substitutions are topics for the discussion section of a paper, whereas in the results section data should be presented without such interpretation. Exploring exhaustively the layers of structure in animal societies is a major undertaking. Not least because a major objective of studying social dynamics is to contribute to the solution of practical conservation problems, it is inevitable that such studies may sometimes have to be undertaken quickly.

The obvious shortcut, within the framework of existing theory, is to use selected revealing behaviour as a guide to an overall system. As a loose analogy, this is akin to using the seating arrangement at a formal dinner as a guide to the social role of the guests in contexts beyond the dinner. Naturally, such shortcuts necessitate validation.

BEHAVIOURAL PARAMETERS

Even to a casual observer it is obvious that most behaviour patterns occur in bouts; that is, they do not occur randomly in time. Although the existence of bouts may be obvious, how best to define them is not. A review of major textbooks and many papers reveals that many prefer to avoid this issue.

The most common usages define bouts as a repetitive occurrence of the same behavioural act (states or events) or a short sequence of behavioural actions that occur in some functional pattern. States usually have durations, and states with extremely short durations are called events. The difficulties of defining the hierarchy of bouts, states, and events is illustrated by allogrooming by a mouse. This comprises a series of actions (nibbles) of short duration; an uninterrupted string of these nibbling actions might make up a bout of grooming.

However, a student of the detail of mouse grooming will see that these strings of nibbles may sometimes transfer from one body region to another (*e.g.*, from head to neck to flanks or back), and for some purposes it may be helpful to distinguish bouts at this finer scale (a bout of head grooming distinct from one of flank grooming). The problem is that the best definition of a bout depends on the purpose of the analysis in which it will be used.

Depending on the scale of resolution required, even a short sequence of nibbles at one patch on the flank might be distinguished from another bout of grooming at the next patch of fur. Ultimately, each nibble could be defined as a state, punctuated by another state (shifting the head a fraction to grasp the next tuft of fur). At a given level of resolution it may be helpful to define states from which bouts are built up, but often, under closer scrutiny, a state will emerge to have a structure that could itself represent a bout (rather than one nibble, or one sweep of the paws, while grooming).

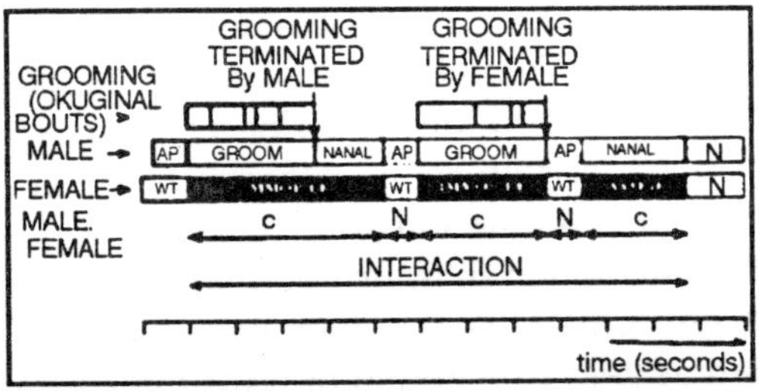

Fig. Data on the Interactions between male and Female Wood mice.

In practice, the best definition of a bout depends on the purpose of the analysis. The top row illustrates that grooming (of the female by the male) is split into a series of brief nibbles partitioned by fleeting changes of position. Each period of nibbling might be defined as an event or as a bout.

If each nibble is an event, then the sequence of them might make up a bout of grooming. Bouts might also be defined in terms of contact between the two mice, and in that case the period during which the male was first grooming and then nasoanal sniffing the females constitutes one bout of contact, during which the female was immobile. Finally, the entire period of male—female interaction might be defined as a bout. This scheme is based on real data on the behaviour of wood mice. Depending on the model under analysis, grooming bouts could be distinguished as one continuous process or as a series of original grooming bouts in which a mouse shifts between body positions. Even an interaction can be considered a bout if several criteria are fulfilled. C = contact behaviour, N = noncontact behaviour, AP = approach, WT = wait, N-anal = nasoanal contact.

In some contexts this wracking down of the microscope to reveal more and more detail may seem merely a quest to bring into view the number of angels perched on the pinhead, but an important point nonetheless emerges: the unit of much behavioural analysis is the bout, and the usefulness of a definition of a bout is affected by the level of magnification at which the analysis is being undertaken. Bouts must be defined very carefully because their definition will have far-reaching statistical consequences for any analysis in which they are involved and because of their role as an indicator of motivation and neural processes. From the mathematical point of view, when a behaviour is modeled it is easiest to keep definitions simple, so Haccou and Meelis define a bout as a "time interval during which a certain act is performed. A bout length is the duration of such a time interval."

In the calculation of transition matrices, transitions to the same act are impossible, so diagonal elements in the matrix are treated as zero (the notion of transitions from one bout of behaviour to another of the same behaviour presupposes that the observer can distinguish a bout from an interval between two consecutive bouts).

The complexities of defining bouts seem particularly difficult to accommodate in studies of mammals, for which one action pattern often appears to slur into another. For example, although the task of defining a bout of drinking by a fox (made up of a string of lapping events) is feasible, the parallel quest of distinguishing the gulps of a rat or primate (which drink continuously) might be dubious. In comparison it would be straightforward, in a parallel study of a bird, to identify consecutive pecks and the interval between them, which can be analysed using log—survivor plots. The awkward truth is that the convenient hierarchy of bout, state, and event is often arbitrary and often concerns a continuum of lengths (clearly revealed by techniques such as the Noldus videotape analysis system, which measures even the durations of events).

To continue with the example of allogrooming wood mice and opting for Haccou and Meelis's definition of a bout as either the duration of a state or the interval between two states, it can be straightforward to recognize bouts. Wood mice, for example, often switch between two states during exploration: scanning while walking and rearing up or scanning while immobile. In this example, the behavioural elements scan, rear up, and walk each have durations and are therefore states. For example, the duration of scanning is a bout. During scanning, however, wood mice often turn their heads back for only a tenth of a second; turn-head-back is an event. In this case, one bout may include scanning interrupted by several turn-head-back events.

Sometimes, however, it is difficult to distinguish between bouts, although methods are available to do so. When behaviour is studied in sequences (*i.e.*, continuous records), the existence of bouts can be confirmed because the distribution of their lengths follows an exponential distribution as long as successive bouts comply with the assumptions of a first-order continuous-time Markov chain model. In practice, if bout durations do not follow an exponential distribution, there are two possible explanations.

First, the observer incorrectly recognized the bouts and therefore measured the wrong thing, perhaps because some bouts were only partially observed through insufficient time for observation.

Second, the first-order Markovian assumption is not upheld. In the latter case, there are again two possibilities. First, bout lengths may exhibit dependency (between successive bouts or every second or third bout). In this case it is always better to use a semi-Markovian model. Second, and more abstruse, there may be second- or third-order Markov models, which according to Haccou and Meelis are rare and intractable in ethology.

In general, bouts approximate to a first-order Markov process and therefore can be distinguished on the basis of testing their distribution or fitting a nonlinear curve to the logarithm of observed frequencies of gap length. Because most behavioural studies are based on parameters such as latency, duration, and time intervals, bout length is a very important parameter to study at the outset.

STATIONARITY

Obviously, if circumstances change over a series of bouts it is confusing or misleading to lump them for analysis. Consequently, it is important to identify stationarity, during which things do not change. Plots of transition frequencies from one act to another in several consecutive periods can be used to judge whether the behavioural process is approximately stationary. Haccou and Meelis emphasize that nonstationarity can mask treatment effects and make the results of an analysis ambiguous. Quite apart from the mathematical implications of inhomogeneity in the data, the existence of the motivational change that causes nonstationarity may be the very object of study, so it is important to identify it from an ethological standpoint. The complexities of stationarity reverberate through ethological methodology. Bekoff, for example, is skeptical that the concept of stationarity can be applied in a social context and therefore concludes that Markovian analysis should be avoided.

THE ETHOGRAM

Early in any study a researcher must classify the behaviour patterns to be documented. This classification is an ethogram, and constitutes a dictionary of the researcher's language; without an ethogram, meaning is not fixed. Ethograms have become unpopular because (like dictionaries) they take up large amounts of space and are dull to read. However, they are vital reference material for those wishing to assess a study's conclusions critically. As a corollary, it is not uncommon for the brunt of reviewers' comments now to fall on statistics rather than data acquisition, reflecting the same shift in emphasis away from data, and onto interpretation, as the hard currency of behavioural science. By analogy, consider the shift from real objects to paper representations in financial systems: 50 years on, would you rather find a stash of gold or war bonds?

We propose that the Internet, with its capacity to archive and disseminate large quantities of multimedia information, will revolutionize the use and usefulness of ethograms. For example, in parallel to a written publication on badger vocalizations, we have created a multimedia vocal ethogram and another based on digital video. Ethograms are often constructed in a pretrial period during which behaviour appropriate to the study is chosen and described. However, short studies are unlikely to detect a complete catalog of relevant behaviour and are likely to miss social behaviour characteristic of context-sensitive dyadic interactions. Fagen tentatively proposed that a sample of 50,000 acts or more might generally be necessary to estimate the repertoire of a typical carnivore or primate species.

Fagen's estimate was based on the biologically general type—token relationship. This relationship was shown by May to be a general statistical property of all catalogs in which an infinite number of objects belong to a finite number of categories. It prescribes that the logarithm of the number of types in an inventory depends approximately linearly on the logarithm of the total number of acts in the catalog. In practice, our badger study confirmed the point: Some antagonistic behaviours (coordinated attack on the rump of an animal after a partner has grasped its head) that we predicted might occur from their occasional appearance in ritualized play between cubs were seen for the first time in earnest between adults only after thousands of hours of direct observation. Although rare in badgers, these coordinated attacks gave an important insight into a previously undocumented realm of cooperative aggression.

What are the components of an ethogram? Ideally, each element might be purely descriptive, free of any imputed function. A heroic attempt at this purity of description was Golani's use of balletic choreographic scores to quantify the spatial and sequential organization of body part movement, together with its qualities such as speed and force and the degrees of variation tolerated within categories. It was intended to help solve what Golani called a blind spot in the behavioural sciences: the need for a universal language to describe animal movement.

Despite potential for universality of description, such quantitative choreography has proved impractical for most field studies. Also, focusing on the minutiae of postures may be too detailed for this purpose. A realistic option, facilitated by the Internet, is the creation of an archive of film clips and spectrographic catalogs within a taxonomic library. However, such a "content and quality" ethogram is still denuded of context. Is a bird pecking a conspecific attempting fighting, grooming, or feeding? The answer may seem obvious from the rate of pecking, its strength, and the reaction of the other bird, but logging these contextual clues is time-consuming.

This is why many ethograms incorporate a shorthand notation that uses not only purely descriptive language but also proximate function language, derived from context and likely result. The demons of tautology and teleology lurk in such language.

BEWARE TELEOLOGY

Omnipresent dangers in the study of social behaviour are the closely allied traps of teleology and unwitting anthropomorphism. Teleology, or the doctrine of final causes, infers purpose in nature (*e.g.*, to infer the existence of a creator from the works of creation). The teleological conviction that mind and will are the cause of all things in nature is not within the scope of scientific method. In the context of animal behaviour, teleological explanation would name and account for a behaviour by its presumed ultimate effect (*e.g.*, appeasement, submission, or punishment) and not by its proximate causes or physical appearance.

It may be convenient to label as "punishment" the category of attack launched by a dominant meerkat on the only member of her group not to join in a fight with territorial trespassers. The danger lies in (inadvertently) interpreting the functional nuance of this convenient label as the proximate cause of the attacker's behaviour. We know only that one individual did not join the fight and that the dominant member of its clan then initiated an attack on it (joined by all its group-mates). It is a matter of interpretation whether this attack functioned as punishment and a matter of speculation whether the attacking meerkat had punishment (or anything else) in mind when it attacked; it is certainly unwarranted to conclude that the putatively punitive attack was launched with the ultimate purpose of increasing the fitness of the attacker (although that may well be its consequence).

Interpretation based on context again harbors the pitfalls of premature use of proximate function language. It would be folly, for example, to categorize as "supportive" an instance of a large female grooming subordinate kin but to label as "repressive" the grooming by the same female of nonkin subordinates. In such a case, proximate function language would have prematurely slipped all the way to ultimate function language. Clearly, the risk is that the very hypothesis used to make this interpretation may later be said to be supported by the observation; in this hypothetical case, the erroneously circular conclusion would be that kin selection theory is supported by the observation that females are supportive of kin and repressive to nonkin. In fact, such theories can be tested only by following actions through to fitness consequences—a very difficult task. In the progress of a given scientific inquiry, there is no virtue in premature exchange of data language for theory language. The teleological trap arises because in humans, conscious action (and even empathy to other people's unconscious actions) often involves mental plans and objectives that stimulate or explain the performance of behaviours. But consciousness has not been proved in other species and to confuse proximate factors with ultimate factors in this way may stifle correct interpretation.

As Kennedy noted, "Teleological terms such as 'searching' ... really describe the animal's presumed state of mind. ... Trying to find an objective substitute for a teleological term will always pay off in research because it forces a mental break-out and a closer look at the components of the behaviour actually observed, components which the handy teleological term leaves unnoticed or at best unformulated in the back of the mind." Kennedy concluded that scanning would be a more useful term for searching. The worst cases of teleology may use such "catchy" labels that proximate and even ultimate function are obscured in the mind of the reader (if not the writer) because of the human innuendo of the word. Rape, used as a label for resisted matings, has been widely considered a case in point.

Used correctly, "mock anthropomorphism" can be a very valuable heuristic tool to guess the function of behaviour, using our own mental processes to explain behaviour in an ultimate fitness context. Such hypotheses can be tested, but even if they are supported we must beware of unwittingly crossing the threshold of assuming that the animal has used the same mental processes when deciding on its own actions. The temptation of teleology—to impose our mental model on others—may be greatest when dealing with other primates. Tinbergen foresaw this when he wrote, "Teleology may be a stumbling block to causal analysis in its less obvious forms. ... The more complex the behaviour systems we deal with, the more dangerous this can be." These strictures against teleology do not imply that ethologists diminish animals to the level of

Descartean machina anima; rather, the objective is to be mindful of Lloyd Morgan's canon that. In no case may we interpret an action as the outcome of the exercise of a higher psychical faculty, if it can be interpreted as the outcome of the exercise of one which stands lower in the psychological scale". This is a methodological rather than ideological stance, which may be more likely to be correct than the converse assessment. As Dawkins put it, a pragmatic aid in navigating this tricky terrain is to advance "always reassuring ourselves that we could translate our sloppy language into respectable terms if we wanted to."

CLASSIFICATIONS OF BEHAVIOURAL INTERACTIONS

How, in practice, can we impartially assign positive or negative implications to behavioural interactions such as proximity or grooming that are ambivalent, without recourse to the very theories we seek to support? One solution lies in sequence analysis, using unequivocal behaviour patterns as anchors; high-intensity attacks are certainly detrimental to the receiver's fitness, whereas allowing mating is usually beneficial to fitness. If transition probabilities revealed grooming as a predictor of attack, then it might usefully be classed as aggressive, but if grooming often precedes acceptance of mating, it is probably amicable. Firm grounding of a behaviour in context may justify an interpretative classification. Often, however, it is impossible to establish a uniform link between ambivalent behaviours (*e.g.*, grooming) and anchors (*e.g.*, coalitionary aid or attack) until the whole social network of interactions within the group has been described. Descriptions based on ultimate function should be eschewed until such links are established.

A very long list of behaviour patterns or relational types could be described and analysed at each level of the structural hierarchy of social dynamics. Here, we consider just one example at each of three levels: grooming as an interaction, dominance as a relationship, and the concept of "social group" as a structure.

- *Grooming*: Allogrooming can be abundant, observable, and clearly bidirectional or unidirectional, a combination of attributes that have made it perhaps the most frequently studied social interaction. Despite the common assumption that it is amicable or beneficent, it may also cue either dominance or submission in a ritualized context.

The style of grooming, be it simultaneous, alternating, or routinely one-sided, is often largely species typical and even a small sample may give important clues to social relations in a wider context. Grooming has been the focus of many primatological studies, among which it appears to have more to do with social bonding than with hygiene. For example, there is a significant correlation between time spent grooming and group size but not body size. The idea that primates compete for grooming access in a social setting has also been very influential.

To unravel more complex intricacies requires more probing analytical techniques, of which an especially rigorous example is Hemelrijk and Ek's study of reciprocity and interchange of grooming and agonistic "support" during conflict in captive chimpanzees. Reciprocity was defined as one act being exchanged for the same thing (grooming for grooming), whereas interchange was defined as involving two different kinds of acts being bartered (*i.e.*, grooming for support). They created an actor matrix (who initiates to whom) and a receiver matrix (who is recipient from whom) and used a correlation procedure, the K r test, to examine links between the two matrices while taking account of individual variation in tendency to direct acts.

They also attempted to retain stationarity in the data by distinguishing between periods when an alpha male was clearly established or when alpha status was in dispute. Hemelrijk had shown that grooming and support were both independently correlated with dominance rank, so whenever reciprocity or interchange was part of a triangle of significant correlations, they partialed out the third variable to see whether the remaining two retained significance. Between the females, for example, a significant interchange between grooming and support received did not depend on dominance rank. Females groomed individuals that had supported them regardless of rank, indicating a social bond. However, the reverse was not true in that support was not given in return for grooming, so it was not possible to "buy" support with grooming. Patterns of relationships were different between the sexes. Cheney used the distribution of grooming among individuals within a group to examine whether adversity, in the form of rivalry between groups, had a uniting effect within the group; the expectation of this hypothesis was a more egalitarian spread of grooming within the group when competition with neighbours was intense.

This expectation was not upheld, perhaps because neglected individuals would suffer a greater loss of fitness by shirking in their support of their own group than by attempting to trade this support for better treatment, even assuming a lack of punishment for failure to cooperate. Even when intergroup competition was strong, marked hierarchies in grooming favouritism characterized intragroup relations. This study used the distribution of grooming to generate data with which to infer intragroup cohesion and equality. The ultimate functional explanations given depend directly on the validity of that assumption.

Our acceptance of such hypotheses may be influenced more by our intuitive anthropomorphic bias for their seeming correctness than by our purely objective assessment of the supporting evidence. This is the dilemma of being a social mammal attempting to describe the societies of other mammals. The site to which grooming is directed may have social implications. Among long-tailed macaques (Macaca fascicularis), for example, low-ranking females expose the chest, face, and belly less often to higher-ranking females than vice versa. Male—male dyads almost never groomed the face, chest, and belly. Similarly the face, chest, and belly were underrepresented in grooming between individuals that groomed each other less frequently.

One interpretation is that recipients of grooming expose less vulnerable parts of their body to, and avoid eye contact with, individuals they perceive as potentially dangerous.

Another analysis of the form, distribution, and context of social grooming was undertaken in two groups of Tonkean macaques (Macaca tonkeana) by Thierry *et al*. They found that kinship and dominance role had no effect on the form or distribution of social grooming among adult females, the most common class of individuals to be observed grooming. The caveat that no one method of disentangling social relationships is universally appropriate has been stressed by Dunbar. For example, in a study of grooming in three species of macaques, Schino *et al*. found that whereas frequency and total duration of grooming were highly correlated, there was low correlation between its total duration and mean duration, and frequency and mean duration were not correlated.

A closer inspection of the data also revealed that even this general equivalence between measures of frequency and total duration did not always hold within specific grooming relationships. In general, frequency is a measure of initiation and mean duration is a measure of continuation. In Schino *et al.*'s case, both measures were needed to interpret adequately the macaques' society. Not surprisingly, one important stimulus for grooming in rodents is soiled fur. However, because grooming can be undertaken alone or mutually, it has social implications transcending cleanliness.

Stopka and Macdonald (in press) used a similar approach to Hemelrijk and Ek's analysis of grooming in captive chimpanzees to examine patterns of reciprocity in grooming among wood mice (Apodemus sylvaticus). A powerful method for investigating this involved row-wise matrix correlation and the Mantel test. Allogrooming was less common than autogrooming and was most commonly directed by male wood mice to females. Having described a social network in grooming, Markov chain analysis then revealed that the termination rate of female contact behaviour (expressed as her tendency to flee the male's attentions) depended largely on the male's tendency to terminate allogrooming. The pattern of transition rates from one behaviour to another revealed that the most common transition was from allogrooming of the female by the male to male nasoanal contact with the female.

This indicates the motivation underlying the male's tendency to allogroom females: males groom a female in order to gain the opportunity to sniff her nasoanal region and, ultimately, to decrease her tendency to flee, thereby allowing them the opportunity to mate.

Females are most likely to mate with males that groom them the most. This constitutes an unusual example of the market effect in a short-lived mammal, with females essentially selling sex in exchange for grooming. Allogrooming is the clearest and most common form of cooperation among badgers. It appears to be a valuable supplement to self-grooming because it is focused on the parts of a badger (back of neck, shoulders, and upper rump) that are most difficult for an individual to reach for itself. The proportion of grooming directed to each segment of the body differs, but is complementary between self-grooming and allogrooming.

Allogrooming cooperation is maintained in this case by the simple expedient of ensuring that as much grooming is given as received. Adult badgers groom one another simultaneously using a responsive rule set that dictates that grooming can be initiated generously, but rapidly withdrawn (at a mean of just 1.2 seconds) if the recipient does not respond in kind. It shows that although self-grooming or allogrooming alone is strongly biased towards particular body regions, a combination of self-grooming and allogrooming produces a fairly even body coverage. This supports a generally utilitarian function for badger grooming.

10

Declining Amphibian Populations

Declines have been noted in diverse habitats including areas far removed from known sources of human interference. The pattern is puzzling. Populations of certain species are dwindling even as sympatric species are thriving. In some areas virtually all frogs have been wiped out; yet elsewhere few if any species appear to have been affected. The specter of mass extinctions within an entire vertebrate class is ominous not only because amphibians are key components of many ecosystems but because whatever factors are responsible might also be affecting other vertebrates including ourselves.

The scale of the problem has been realised only recently because few long-term studies exist, fluctuations in amphibian populations are not unusual, and effects vary tremendously among species. Anecdotal accounts are much more numerous than longterm quantitative data, but available data support the hypothesis that species worldwide are declining in synchrony. Amphibian populations regularly vary in size from year to year. These oscillations can be dramatic both in magnitude and frequency or more subtle and difficult to detect. During the past 40 years, major crashes of frog populations periodically have been reported, but often they have been followed by remarkable recoveries. The large number of concurrent reports now emerging suggests, however, that these events are occurring with increasing frequency to produce an observable overall trend.

No single factor or group of factors can account for the global population declines that we are witnessing. Causation needs to be analysed on a case-by-case basis. Some previously common amphibians have dropped in numbers simply because they have been overcollected and others have disappeared as their habitat has been destroyed but most declines remain mysterious. Often, perhaps no individual factor is responsible, but factors act synergistically to increase mortality and ultimately produce observable declines. We believe that amphibian declines can be understood only by examining first how individuals respond to environmental stresses, and next how they interact with one another in populations facing these stresses. Behavioural ecology can provide insight into the dynamics of declining populations and can provide methodology to identify and monitor populations most at risk.

CAUSES OF POPULATION DECLINES

Numerous factors have been suggested as potential contributors to amphibian declines. Primary suspects include both physical and biotic environmental changes. Susceptible individuals may be directly affected, or environmental changes may induce shifts in population structure or community dynamics that ultimately lead to declines. As populations decrease in size, their vulnerability to stochastic events increases and they may be drawn into "extinction vortices" in which environmental, demographic, and genetic factors reinforce one another to hasten declines. In this synopsis, we highlight current controversies and areas under active investigation.

CHANGES IN THE PHYSICAL ENVIRONMENT

Increased transmission of ultraviolet radiation through the atmosphere due to depletion of stratospheric ozone may threaten amphibian populations. Significant increases in levels of UV radiation have been detected in many areas but populations at extreme southern latitudes and those at higher altitudes may be subjected to especially high doses. UV-B radiation can cause potentially lethal damage to developing amphibian embryos, larvae, and metamorphosing individuals. The idea that amphibian declines may be linked to the thinning ozone layer has recently led to experimental tests of effects of UV-B exposure or survivorship of eggs and larvae. Embryos of at least one species of frog (*Rana cascadae*), toad (*Bufo boreas*), and salamander (*Ambystoma gracile*) show increased mortality when exposed to natural levels of UV-B as compared with embryos covered with UV-B filters; but other frogs appear unaffected. Blaustein *et al.* hypothesised that species that typically spawn in sites exposed to sunlight have evolved efficient DNA repair mechanisms and found, consistent with this, higher levels of key repair enzymes in *P. regilla* and *R. aurora* than in the other amphibians tested.

The UV-B findings have been widely reported in the popular media as "the first to directly link a specific environmental cause, other than habitat loss, to reports on scattered worldwide declines and even local extinctions of some frogs, toads, and salamanders". Many details are lacking from the published experimental methods, however, and the results and interpretations are controversial. The possibility that UV-B affected embryos indirectly, for example by leaching toxic solutes from experimental enclosures, was not considered.

Moreover, the jelly coats surrounding amphibian eggs serve as natural UV-B filters; by removing some layers, the researchers may have inadvertently biased their results. Egg jelly rapidly swells from the time of oviposition, and this may provide considerable protection after a few hours. Because many amphibians breed at night, newly laid eggs might be safe from UV-B damage during their most vulnerable period. UV-B radiation fails to penetrate even 1 cm beneath the surface of many ponds because dissolved organic matter, algae, and waterborne particles effectively block it Licht and Grant. Most amphibians thus would be protected, but those that breed in shallow, high-altitude mountain ponds with clear water, such as the newt *Triturus alpestris*, may be vulnerable.

Many species declines cannot be explained by UV-B effects. For example, the Australian hylid frogs *Litoria aurea* and *L. raniformis* have experienced massive range contractions during recent years, while populations farther south in New

Zealand, which have experienced larger increases in ambient UV-B radiation continue to thrive. Indeed, survival to the larval stage appears to be unaffected by UV-B radiation both in *Litoria aurea* and its sympatric congeners whose populations are stable.

Certainly we need to continue to evaluate UV-B radiation as a potential cause of amphibian declines, especially in the Southern Hemisphere where UV-B radiation flux is highest in the spring and summer commensurate with the breeding season of many amphibians. In contrast, UV-B exposure in the North Hemisphere has increased most in the autumn and winter when few amphibians are breeding. UV-B is unlikely to explain declines in recent years of frogs and toads in tropical and subtropical regions (*e.g.*, montane populations of Australian rainforest frogs) because UV-B levels at these latitudes should not appreciably differ now from 15 years ago.

CLIMATE CHANGE

On a global scale, climate change might conceivably explain simultaneous changes in distant populations. Even short-term variation in weather patterns can dramatically affect amphibian populations. Amphibians need moist habitats to live and reproduce, so desiccation is an incessant threat. The abundance and composition of the invertebrate prey in their diet is likely to shift with changing environmental conditions. Reproductive behaviours, spatial distributions, and vulnerability to predation and disease are likely to change as conditions become warmer or drier.

Some recent population declines coincide with changing weather patterns. Persistent droughts may have resulted in extinction of populations of leopard frogs Rana pipiens in North America and corroboree frogs *Pseudophryne corroboree* in Australia. Altered weather patterns may be responsible for wide-ranging declines of frog populations in Brazil and the possible extinction of the golden toad *Bufo periglenes* in Costa Rica found that the survivorship of adult wood frogs *Rana sylvatica* declined in months with reduced rainfall, and Semlitsch *et al.* documented a close relationship between rainfall and breeding success in numerous species over a 16-year period. Prolonged dry spells led to massive die-offs of adult and juvenile *Eleutherodactylus coqui*. Then as populations began to recover, they disappeared following severe damage inflected by a hurricane, only to reappear as the forest recovered. But many amphibian declines cannot be adequately explained by changing weather patterns.

ACID RAIN

Amphibians have complex life cycles, typically with aquatic larvae that metamorphose into adults that live on land or in water. Degradation of either larval or adult habitat can have serious effects on populations. Acid rain was suggested early on as a serious threat to amphibian populations because of its devastating effects on larval survivorship.

Acidic pulses resulting from snowmelt dramatically alter the chemistry of temporary ponds in which many amphibians breed and, when these pulses coincide with highly vulnerable early embryonic stages, significant mortality occurs. The natural acidification of soils also may be accelerated, causing mortality or reproductive failure in terrestrial amphibians. Developmental aberrations and decreased larval growth rates have been noted in acidified environments. In recent times, many breeding habitats in eastern North America Britain and Scandinavia have become sufficiently acidified to cause breeding failures.

Yet selection favours acid tolerance, and some species persist in acidified conditions, although surviving individuals may be impaired. In other regions, such as in western North America where numerous species are in severe decline, acid rain normally does not appear to be a problem. Dunson *et al.* caution that the loss of even a single population as a result of acid rain has yet to be rigorously documented.

Effects of acidification on amphibian populations vary depending on many factors, including the presence or absence of toxic metals, dissolved organic matter, and the algal communities present. Some amphibians may detect acidified conditions prior to spawning, and then search for more suitable localities in which to breed, but individuals inhabiting acidified ponds apparently lack these behavioural discrimination abilities.

PESTICIDES, HERBICIDES AND FERTILIZERS

Their complex life cycle also makes amphibians especially vulnerable to toxic chemicals such as fertilizers pesticides and herbicides used in agriculture. Early embryonic stages are extremely sensitive to environmental insults. Mortality may result from exposure to agricultural chemicals and industrial pollutants. Fragmentation contributes to declines in two ways: by altering the demography and by modifying the genetic structure off populations.

DEMOGRAPHIC EFFECTS

Local populations are normally interconnected by the migration of individuals among them. The network comprises a metapopulation. Migration into peripheral populations may be essential to prevent their extinction. Moreover, when one population goes extinct within a metapopulation, another should eventually be reestablished at the same locality through recolonization.

This may happen rapidly if potential source populations are nearby and if substantial dispersal occurs from them, but may take many generations, or fail to occur at all, when populations are distantly scattered and dispersal rates are low. Recent empirical studies show that frog populations respond as theory predicts: the probability that local populations go extinct increases in proportion to the distance between populations. Barriers further occlude migration and can lead to a progressive breakdown of the metapopulation, thereby preventing recolonization and accelerating species declines.

Such efforts have been documented in a variety of frogs. What constitutes a barrier varies among species; although forest-dwelling species may find themselves being constricted into islands by deforestation, for example, other species successfully exploit such disturbed habitat. Metapopulations can collapse to extinction without warning even in habitats that are only gradually degrading. Reports of rapid simultaneous declines of numerous species within a region, even as some species appear to flourish are thus consistent with the predictions of metapopulation theory.

GENETIC EFFECTS

Dispersal among local populations not only ensures their persistence by maintaining stable population sizes but enlarges genetically effective population sizes (Ne). Gene flow builds up a reservoir of genetic variation that may be necessary to generate evolutionary responses to environmental changes. Moreover, both the likelihood and potential costs of inbreeding decrease as population sizes increase.

As barriers arise to gene flow and metapopulations break down, populations become more genetically homogeneous. Less-common alleles are increasingly lost from populations by genetic drift and individuals become more homozygous. Mildly deleterious mutations may become fixed within populations, increasing the risk of extinction.

After localized extinction events, those populations that recover are likely to be initiated by fewer founders, which in turn further reduces levels of genetic variation within populations and increases differentiation among populations. Consistent with this model, Reh and Seitz found little genetic polymorphism in common frog Rana temporaria populations isolated by motorways or railroad lines, and *Sjögren* found especially low genetic variation in peripheral populations of pool frogs *Rana lessonae.* The incidence of breeding among close relatives is likely to increase. Deleterious genetic effects then result from the expression of recessive deleterious alleles or overdominance.

In ectothermic vertebrates, hatching success, growth rates, developmental stability, size at maturity and courtship behaviours all show evidence of inbreeding depression. Conversely, outbreeding can benefit progeny by conferring on them fitness benefits associated with increased heterozygosity. Heterozygous tiger salamanders *Ambystoma tigrinum* grow faster and have greater metabolic efficiency than homozygous individuals. Survival correlates positively with the level of multilocus heterozygosity in overwintering juvenile western toads *Bufo boreas* and reproductive success of green treefrogs *Hyla cinerea* correlates positively with their heterozygosity.

Genetic variation confers important benefits both on individuals and on populations, especially in stressful environments. Some amphibian populations lack much genetic variation but nonetheless persist over many years with little sign of inbreeding depression in stable habitats. Theory and laboratory studies predict, however, that once a threshold of inbreeding is reached, populations are certain to go extinct, often without any warning.

DISEASE

Population declines sometimes are associated with identifiable bacterial, fungal, or viral pathogens. Epidemics of the bacterium *Aeromonas hydrophila,* known to cause symptoms of "red-leg" (a highly contagious disease especially common in captive populations) may have been responsible for massive die-offs of natural populations of adult American toads *Bufo americanus* larval wood frogs Rana sylvatica larval and adult mountain yellow-legged frogs Rana mucosa and larval and adult midwife toads *Alytes obstetricans.*

Aeromonas hydrophila and other bacteria are commonly isolated from populations of dying frogs, toads, and salamanders. Pathogenic fungal infestations similarly have been linked to mortality of embryos and larvae in populations of various frogs and toads. Fungi may be the primary agents of decline in some species, and they in turn increase the vulnerability of their hosts to secondary bacterial infections. The geometry of egg-laying can create conditions that promote fungal infestation.

Species that deposit their eggs communally appear more at risk than those that scatter their eggs, presumably because fungi are more readily transmitted over close distances. Moreover, aberrant oviposition behaviour, which may be induced by environmental changes, can exacerbate these effects -for example, rather than wrapping their eggs in strings around vegetation as toads typically do, *Bufo bufo* in some localities have been observed to lay their eggs in compact clumps from which no embryos survived. Although rapid fungal infestation is commonly associated with the mortality of clumped embryos, asphyxiation due to insufficient diffusion of oxygen through the gelatinous egg mass needs to be considered as a possible cause of death. Oxygen depletion of the water due to eutrophication is likely to compound this problem.

Patterns of population declines ("extinction waves") in Australia, Central America, and the western United States are typical of epidemics involving highly virulent infectious agents. Iridoviruses isolated from dying frogs in Australia and elsewheredies have been targeted as a likely cause, and Laurance *et al.* speculate that these viruses have been spread worldwide by exotic aquarium fishes. Parasites also are known to cause epidemics in captive amphibian populations and apparently can cause rapid declines in wild populations. Unidentified protozoan parasites have been found in the skin of frogs succumbing to epidemics in remote regions of Panama and Australia.

Skin lesions caused by these parasites ultimately may kill frogs by interfering with their respiration and osmoregulation. Parasites also may be responsible for the large number of deformities recently observed in anuran tadpoles in central and eastern North America although pesticides or other agricultural chemicals may be implicated in some cases.

Healthy individuals may remain healthy even in the presence of pathogens because of their immunological competence. Immune systems, when weakened due to environmental stress or viral action, potentially leave individuals vulnerable to bacteria, fungi, or parasites from which they are normally shielded and with which under ordinary circumstances they may live commensally.

SYNERGISTIC INTERACTIONS

The causes of amphibian declines defy simple explanations probably because complex interactions among environmental stresses produce effects of greater magnitude than are revealed in investigations of single factors alone. Environmental stresses break down defences against pathogens and parasites.

UV-B, for example, can increase susceptibility to viral, bacterial, fungal, and parasitic infections. Long *et al.* found that, although neither acidic pH nor UV-B alone influenced the survival of leopard frog *Rana pipiens* embryos, embryonic survival was reduced under conditions of simultaneous exposure to both acidic pH and UV-B.

Similarly, pathogenic fungi more readily attack embryos already weakened by exposure to cold temperature, acidic pH or UV-B radiation. Many environmental pollutants increase the susceptibility of amphibians to disease. Climate change can produce behavioural and ecological changes that increase disease risk.

One possible pathway for these synergistic effects is through the induction of immunosuppression. Environmental changes that appear harmless nonetheless may cause sublethal stress, which in turn increases the secretion of adrenocortical hormones that can reduce the animal's ability to fight disease. Even handling frogs momentarily causes elevated corticosterone levels over extended durations so just by studying declining populations we may be stressing them and hastening their demise. Corticosterone also effectively reduces circulating levels of sex hormones, thereby inhibiting courtship behaviours. The cumulative effects of environmental stresses on endocrine and immune system function can result in increased mortality, decreased reproductive rates, and ultimately population declines. Whether declining amphibian populations suffer immunosuppression—attributable to low pH, cold temperature, or other factors—needs to be addressed with standard immunological methods.

Comparisons of corticosterone levels between declining and nondeclining populations have yet to be made, but results would be difficult to interpret. Corticosterone causes a decrease in circulating lymphocytes but also a transient increase in circulating heterophils, which are primary defences against bacteria. Moreover, corticosteroids mediate many physiological processes, and high levels do not always indicate impaired immune function.

The most puzzling extinctions worldwide have occurred in montane populations. Because amphibians are ectothermic, their body temperatures correspond to the cold ambient temperatures typical there. Low body temperatures can induce immunosuppression. In this state, amphibians may become susceptible to pathogens or parasites to which they are normally immune; indeed, *Aeromonas hydrophila* and other bacteria associated with dieoffs can be found in and on healthy individuals. Available evidence suggests that many environmental changes (shifts in temperature, water balance, UV radiation, and water chemistry) represent stresses that make animals vulnerable to disease. When pathogens increase in number and virulence, they can reach a critical level at which even healthy individuals become susceptible and epidemics may ensue.

Disease risk also rises as genetic homogeneity increases within populations. To the extent that species inhabiting restricted montane habitats consists of genetically isolated populations they may be more vulnerable to disease. Genetically similar individuals share similar immunological defence systems. When a pathogen overcomes one individual, it likely will overcome others as well. Thus the rate at which pathogens spread, and whether they can build up to a critical level in the population, is likely to depend on the degree of uniformity among the host genotypes present. In a similar way, individuals within genetically homogeneous populations are likely to show correlated responses to environmental peturbations, increasing the risk of cataclysmic declines.

Effects of inbreeding can accumulate over time within populations gradually decreasing reproductive rates and increasing susceptibility to disease, parasites, predators, competitors, and climatic changes. Extinctions of inbred populations may be triggered by demographic or environmental causes, yet ultimately result from the cumulative effects of inbreeding especially in changing environments.

DISPERSAL, NATAL PHILOPATRY, AND GENETIC POPULATION STRUCTURE

Studies of the behavioural ecology of amphibians suggest that genetic homogeneity poses special problems for them. Amphibians show remarkable site fidelity. Many frogs toads and salamanders migrate year after year to particular sites to breed—even if the ponds are paved over into parking lots.

This site consistency suggests that amphibians may not readily recolonize extirpated populations and efforts of environmental planners to move breeding populations even short distances often fail. Inferences drawn from adult behaviour may be deceptive, however, as significant proportions of juveniles have been found to disperse in those frog and toad populations that have been surveyed in longitudinal studies.

Breden marked Fowler's toads *Bufo woodhousei fowleri* as larvae or at metamorphosis and found that most returned to breed in their natal ponds, but some dispersed to nearby ponds, and a few migrated over long distances. Most wood frogs *Rana sylvatica* also demonstrate natal philopatry, but a minority regularly disperse usually nearby but sometimes to ponds up to 2.5 km away.

Although adult pool frogs *Rana lessonae* show extreme site fidelity, more than a third of each juvenile cohort emigrates from natal ponds. Similarly, most common toads *Bufo bufo* return to their natal pond to breed, but one out of six disperses to other ponds. Although juvenile migration normally ensures some gene flow between local population the large proportion of philopatric individuals is likely to give rise to genetic differentiation among populations, especially when populations are small. Genetic substructuring increases the vulnerability of amphibians to declines in two ways.

First, because relatively few individuals disperse even in undisturbed habitat, metapopulation structure can be easily disrupted by fragmentation.

Second, high levels of genetic homogeneity within local populations, even in undisturbed habitat, can lead to inbreeding depression and increased risk of disease. As possibilities for dispersal are reduced, fragmentation leads ultimately to even further depletion of genetic variation and higher levels of inbreeding. To the extent that inbreeding decreases fitness or genetic homogeneity increases disease risk, fragmentation creates a positive feedback loop that escalates the probability of population extinction with each generation.

Estimates of dispersal typically have been determined with mark-recapture studies. These present numerous logistical challenges and they tend to overestimate levels of gene flow. Genetic population structure can be determined more readily by molecular analyses. Because protein polymorphisms usually vary only over larger geographic ranges few studies have examined genetic differentiation among nearby amphibian populations.

Driscoll *et al.* found significant genetic subdivision among populations of the threatened Australian frog *Geocrinia alba,* both within and among creeks, over distances of 5 km. Genetic subdivision of local populations of the common frog *Rana temporaria* was found only where localities were surrounded by substantial barriers such as motorways and railroad lines. Analyses of nuclear DNA variation among declining populations of the leopard frog, *Rana pipiens*, reveal some differentiation over 3.5 km.

Similar analyses that we currently are conducting on threatened populations of the New By examining variation in mitochondrial DNA (mtDNA), Waldman *et al.* found striking levels of genetic differentiation among local breeding populations of American toads (*Bufo americanus*). During most of the year, toads in these study populations inhabit a wooded reserve. During the spring, however, individuals migrate, some over distances of several kilometers, to ponds at the edge of the reserve.

There they typically remain for several days before breeding and then disperse. Although the ponds are in close proximity to one another (distances range from 0.8 to 2.2 km), mitochondrial DNA haplotypes of individuals mating in these localities tend to differ. Genetic similarities within ponds, and differences among ponds, persist year after year, suggesting strong natal philopatry of females. Nuclear DNA fingerprinting also points to large numbers of male progeny returning to natal sites to breed. Although *Bufo americanus* populations fluctuate in size from year to year the species readily exploits disturbed habitat and appears to be thriving. Even had we not studied these toad populations over many years, the substantial genetic variation apparent within populations would have suggested that the species was not in decline.

INBREEDING AVOIDANCE

The typical amphibian population structure is conducive to high levels of inbreeding. Some individuals disperse and are unlikely to inbreed. But most individuals—of both sexes—return to natal localities to breed. Once there, they encounter siblings and other close relatives as potential mates. Dispersal does not provide a reliable means of inbreeding avoidance for amphibians as it does for many other vertebrates. In declining populations, the frequency of mating between close relatives is likely to increase. If amphibians can recognize their close kin, they might avoid mating with them.

Over many years, we have studied natural populations of *Bufo americanus* for evidence of inbreeding and possible mate choice. Toads are easy to observe, perhaps because they are poisonous and not easily intimidated by potential predators, and they appear to behave normally even when approached or handled. Thus we were able to employ both observational and experimental approaches to estimate the frequency of inbreeding and to analyse mechanisms of mate choice.

Upon arrival at breeding ponds each spring, male toads produce trilled calls which can be easily heard up to 1 km away. These probably serve as advertisement signals for females and may attract conspecific males. At any given time, more males than females are typically visible in a breeding aggregation. Males scramble for females, and indiscriminately clasp males and females, as well as other similarly-sized frogs, and even inanimate objects. Clasped males utter a release call which effects their release. Females usually arrive at ponds unamplexed, and upon entering the water they tend to remain submerged and repeatedly approach males. Females often act as if they are evaluating particular males.

Females swim slowly towards males, lifting their heads slightly out of the water, and finally approach one and initiate amplexus. Females sometimes are clasped while swimming, however, and in this case they have little opportunity to exercise mate choice.

Pairs occasionally are harassed by unpaired males; the unpaired male struggles with the amplexed male for access to the female. Attempts to dislodge amplexed males are rarely successful. Most matings occur within a period of 2-5 days at each site, and pairs may remain in amplexus for 24 h or more before oviposition occurs. Polygyny thus is infrequent.

MATING OF TOADS WITH CLOSE RELATIVES

We analysed variation in both mitochondrial and nuclear DNA to assess the frequency of matings between close relatives in natural conditions and the mechanisms by which mate choice occurs. Mitochondrial DNA is maternally inherited, so siblings, as well as more distant matrilineal relatives, share identical fragment patterns. If individuals differ in mtDNA haplotypes (*i.e.*, mitochondrial genotypes), then barring mutations, they cannot be siblings. The application of mtDNA to studies of population structure and sexual selection would be limited if particular haplotypes predominated, but often this is not the case. In many vertebrates, mtDNA undergoes more rapid evolution than nuclear DNA. Consequently, mitochondrial markers can vary extensively among individuals both within and among populations sometimes approaching hypervariable minisatellite regions of the nuclear genome in their diversity.

We captured toad pairs as they were laying eggs, and we later genetically typed them. We inferred that individuals were not siblings if they differed in their mitochondrial haplotypes (generated with a series of restriction enzymes). Mates with identical haplotypes might be siblings, or they might be related through a more distant female ancestor. Of 86 mated pairs collected, members of only two pairs had identical haplotypes. Randomly generated pairings of males and females present at each pond during each season, however, led to a null expectation of 12 matings between individuals bearing identical haplotypes. Thus, significantly fewer individuals mated with close relatives than would be expected if pairing were random, but the expected frequencies are low. Brother-sister matings appear to be exceedingly rare; even the two pairs whose haplotypes matched may have been only distant relatives.

SIMILARITIES IN VOCALIZATIONS OF RELATED MALES

Behavioural observations suggest that most female toads have an opportunity to choose their mates and that mate choice is based in part on their assessment of advertisement calls broadcast by males. Frogs often choose larger or older males by detecting features of their calls that correlate with these characters. We were curious whether females might avoid mating with their close kin based on information that they could discern from males' calls.

In the field, we recorded calls of 15 males in each of four breeding populations. Additionally, we noted information on their size and environmental factors including temperature. Later we genetically typed each male by obtaining nuclear DNA fingerprints of each using multilocus minisatellite probes. We estimated the genetic similarity of calling males within ponds by the proportion of bands they shared and determined proportions of bands shared by siblings and half-siblings by controlled crosses in the laboratory. Next we analysed similarities in both temporal and frequency components of males' calls as a function of the callers' genetic similarity and relatedness.

Genetically similar males, including brothers and more distant relatives, do indeed produce similar advertisement vocalizations. In each breeding population, calls of close relatives were similar in their temporal components (*e.g.*, pulse duration, interpulse interval, rise time, and call duration), whereas calls of genetically dissimilar individuals were much more variable.

The effects of genetic similarity on call structure were additional to those of body size and temperature. Females thus might discriminate between close and distant relatives on the basis of hearing their calls and also discriminate between smaller and larger (or older) males based on the same call parameters.

The pitch of the call, or dominant frequency, appears to be a better trait for females to use to assess male size, however, and our study revealed no correlation between dominant frequency and genetic similarity.

FEMALE MATE CHOICE

Kinship information encoded in males' calls might enable females to recognize their close relatives and to avoid mating with them or even to choose an "optimally" related mate. But can females decode this information and use it? To answer this question, we individually tested 29 females in a laboratory arena, observing their responses to recorded calls of two males, alternately broadcast from speakers on either side of the arena.

In each test, both males were from the same pond as the female subject, but they were probably unfamiliar to her because we had recorded the males in the field during a previous year.

Additionally, we matched stimulus calls for call duration and sound intensity. We genetically typed the females, comparing their multilocus DNA fingerprints with those of the males only after the behavioural tests had been completed. Thus the behavioural tests were conducted blindly.

Females can discriminate between relatives and nonrelatives on the basis of their calls. Subjects showed a strong preference to approach the speaker broadcasting the call of the male genetically less similar to themselves.

Only 2 of 29 females preferred the call of the genetically more similar male. Seven females showed no consistent preference for either call, but their DNA fingerprints reveal that they were no more closely related to one male than to the other. Given our observations of pair formation in the field, we interpret these results as evidence of mate selection. Female toads appear able to preferentially choose nonrelatives as mates. We still do not know how females identify call features of their close relatives and nonrelatives. Kin-recognition mechanisms of larval amphibians permit them to discriminate not only kin from nonkin, but also between different classes of relatives (*e.g.*, siblings from half-siblings).

Specific preferences appear to be acquired through the learning of one's own traits or those of close kin during development. Advertisement vocalizations may be genetically determined, but females do not produce these calls, and they probably never have an opportunity to learn their brothers' calls. Mate choice may involve a kin-recognition system in which females evaluate males' calls based on a genetically encoded recognition template.

IMMUNE FUNCTION

Mate choice affects not only levels of inbreeding, but it may also affect immune function and disease resistance. Allelic variation at loci of the major histocompatibility complex (MHC) is thought to play a key role in immunological defence and effectively permits hosts to respond to much more rapidly evolving pathogens. Individuals with heterozygous MHC alleles, if they can recognize and respond to more foreign antigenic epitopes, may be conferred enhanced disease resistance.

Individuals that mate disassortatively at MHC loci avoid deleterious consequences of inbreeding and accrue general benefits of heterozygosity, but especially reduced susceptibility to infectious diseases. Disassortative mating also generates variation among individuals in their MHC types and thus variation in their susceptibility to particular pathogens. This should further boost disease resistance within populations by reducing opportunities for pathogen transmission.

Models predict that populations with little MHC variation should suffer higher incidences of disease than would genetically heterogeneous populations. O'Brien and Evermann hypothesized that population declines in cheetah *Acinonyx jubatus*, black-footed ferret *Mustela nigripes*, bighorn sheep Ovis canadensis, and other species, are attributable to decreased immunological resistance resulting from extreme MHC monomorphism. Yet many mammals appear perfectly healthy despite having few polymorphic MHC alleles and cheetah populations seem to be constrained more by predation than by lack of genetic diversity.

The biochemical and functional similarities between amphibian MHC loci and those of other vertebrates suggest the possibility that mate choice in frogs may result from selection to reduce susceptibility to disease. Natural outbred populations of leopard frogs *Rana pipiens* and bullfrogs *Rana catesbeiana* are sufficiently variable that no two individuals are likely to share exactly the same MHC alleles.

Comparative studies suggest that while MHC alleles regulate immunity in all amphibians, levels of polymorphism vary among species with frogs and toads typically being more variable than salamanders.

We suggest that comparisons of MHC polymorphisms between declining and nondeclining species, and between healthy and diseased conspecific populations, might establish possible genetic correlates that would aid us in understanding why populations decline. More generally, such comparisons might begin to resolve the role of MHC variation in conferring disease resistance.

The evolution of genetically based kin-recognition systems at first appears enigmatic in that extreme polymorphism is necessary at marker loci and kin selection should deplete this variation. Frequency-dependent selection for mate choice or host-parasite interactions appears necessary to sustain variation in the cues used by this type of recognition system. Only MHC loci are sufficiently polymorphic to produce unambiguous markers of overall kinship identity but the polymorphism presumably is driven by selection for effective immune function.

Conversely, genetically based outbreeding mechanisms may have evolved not to reduce homozygosity throughout the genome but particularly to maintain polymorphisms at MHC loci responseble for immune function. Evidence is accumulating that odours associated with histocompatibility genes influence both altruism and mate choice in organisms as diverse as sea squirts, rodents, and humans. Work is currently underway in our laboratory to examine in amphibians the extent to which kin recognition is attributable to MHC variation. If kin recognition is indeed regulated by MHC alleles, depletion of MHC genetic diversity that is likely to accompany population declines would deprive threatened species of an evolved mechanism to avoid inbreeding.

BEHAVIOURAL ECOLOGY AND AMPHIBIAN CONSERVATION

Environmental perturbations do not necessarily directly cause population declines. Rather, in some cases they may precipitate ecological and behavioural changes that in turn increase mortality or decrease recruitment. For example, breeding habitat may be altered because of changes in weather or anthropogenic activities.

To utilize available habitat, amphibians may need to modify their reproductive timing, perhaps breeding later in the year or during shorter periods. Intra- and interspecific competition for calling and oviposition sites may subsequently intensify, with males altering their calling behaviour in response to acoustical interference. Patterns of intra- and interspecific competition among larvae, and their interactions with predators and prey, will change in unpredictable ways.

When suitable habitat becomes restricted, amphibians might be found in denser aggregations, which might be more readily exploited by pathogens and predators. Simultaneously, food resources may decrease in availability due to the altered social structure. Changes in population dynamics and community structure can have profound effects on demography and reproductive strategies that lead to population declines.

Subtle behavioural and ecological effects may offer the first indication that a population is under threat long before declining numbers are noted. The life history of many amphibians is characterized by high larval mortality due to predation and desiccation. Pesticides, herbicides, fertilizers, and other toxicants alter the swimming behaviour of amphibian larvae, making them more vulnerable to predators. UV-B radiation similarly can induce erratic swimming behaviour. Behavioural changes can make larvae easier to capture.

For example, dragonfly larvae prey more efficiently on Hyla cinerea tadpoles exposed to toxic metals than on controls. Rana temporaria tadpoles exposed to pesticides become hyperactive and are selectively preyed on by newts. Changes in habitat selection, feeding behaviour, and pigmentation patterns also can put exposed larvae at increased risk. Even if a potential toxicant is shown to be harmless to amphibians in laboratory studies, through its indirect effects the same agent may prove lethal in nature. Perspectives gained from behavioural ecology are important in interpreting how genetic stochasticity affects declining populations. As populations become smaller, they are more likely to go extinct due to purely random events, and inbreeding increasingly contributes to this risk.

Conservation biologists believed for most of the past two decades that populations with an effective size of at least 500 can maintain sufficient genetic variation to survive. Lande however, recently argued that sustainable populations must be larger by least an order of magnitude (*i.e.*, Ne = 5000) because of the accumulation of mildly deleterious alleles within small populations.

Ne represents the number of individuals in an idealized panmictic population, so censused population sizes usually need to be considerably larger to achieve the requisite genetic variation.

Most amphibian populations have N_e values significantly smaller than either of these threshold values suggesting that all may be destined for extinction. Yet, our finding that amphibians can use behavioural mechanisms to selectively outbreed points to the need to carefully reevaluate such models based on realistic assumptions. The minimum population sizes necessary to counter the effects of genetic drift might be reduced, and the expected longevity of populations lengthened, as a consequence of mechanisms that enforce outbreeding.

When populations diminish substantially in size and number, captive rearing may be attempted to minimize risks of environmental stochasticity. In these conditions, allowing amphibians to choose their mates may be important in reducing deleterious effects of inbreeding and minimizing the risk of disease. Consider the case of the endangered Wyoming toad *Bufo hemiophrys* baxteri.

Until the mid 1970s, this toad was fairly common in the Laramie Basin of Wyoming, but it then declined rapidly and within a decade it was presumed extinct. A single population was discovered in 1987, but the number of remaining individuals decreased from 394 in 1991 to 155 in 1992 due to adult mortality resulting apparently from bacterial and fungal infections. As the species seemed destined for extinction, in 1994 the Wyoming Game and fish department removed all known remaining toads from the wild.

Toads are now being bred in zoos so that tadpoles can be released back into the wild. Our findings suggest that females choose males with which they are genetically compatible as mates. Pairing males and females haphazardly or randomly for breeding, as is currently being done, thus may not be the safest strategy.

Rather, natural conditions should be simulated to allow individuals to choose their mates. Should this fail, genetically dissimilar individuals can be identified and paired with the aid of DNA fingerprinting techniques. With numbers of dwindling, planning matings to maximize genetic variation may seem esoteric. Yet to ignore potential genetic consequences of mate choice could be disastrous for species recovery Programmes. By considering amphibian conservation from the perspective of behavioural ecology, we can gain new insights on the dynamics of population declines and possible solutions to the problem. Yet behavioural ecology offers no panacea, and its incorporation into conservation Programmes presents disadvantages as well. Research into the behavioural ecology of declining populations can be painstaking and time consuming, but the rapid declines that we are witnessing may require quick decisions. Behavioural studies are likely to reveal complex effects that vary under different ecological regimes, so just as with conventional approaches, general answers may be elusive. Most troubling, our studies of threatened populations may subject them to new stresses that hasten their decline.

BEHAVIOURAL TOOLS FOR POPULATION SURVEYS

Male frogs of most but not all species produce calls to advertise to females and often to communicate to other males. These vocalizations are species specific and presumably function to facilitate species recognition. Calls of many species are easily detected from some distance by human listeners. Counting calling males, or numbers of aggregations of calling males, thus can serve as an efficient method to detect species within areas and to estimate relative population densities.

Many frogs are naturally secretive and others live in habitat that is difficult to survey by visual methods. Not only are visual surveys slow and laborious, but in many cases microhabitats may be seriously disrupted or even destroyed in the process of searching for individuals. Tocher's recent use of audio surveys to examine effects of habitat fragmentation on species diversity of central Amazonia frogs illustrates the power of such methodology to expedite research into the causes of amphibian declines.

The advantages of using calling behaviour as a measure of population density may often outweigh its disadvantages. Still, caution needs to be exercised in interpreting results. Zimmerman discusses numerous assumptions underlying this methodology and suggests possible controls (*e.g.*, for observer reliability).

The number of frogs calling in a chorus does not necessarily correspond to the number of frogs present; calling rates tend to decrease as chorus size increases. Two points especially need to be emphasized.

First, the use of vocalizations to monitor either species diversity or population numbers is only valid if one can establish that calling individuals are successfully reproducing.

Second, many frogs are relatively long-lived, living 20 years or more. Because of their site fidelity, frogs may return to call from, and even mate at, particular breeding localities long after the sites have become unsuitable for supporting larvae.

Visual surveys of adults can suffer from the same weaknesses. Population declines of long-lived species may have been underway for years, even decades, before surveys hint at any problems. By then, declines may be irreversible. While the species specificity of frog calls makes them suitable for determining the presence or absence of species, and to some extent the numbers of frogs present, vocalizations encode more specific genetic information as well.

Geographic variation in mating calls of conspecific frogs has been noted for some time. Only recently, however, has such variation been investigated on a more fine grained scale. Substantial differences in call characters between populations of cricket frogs Acris crepitans in different ecological settings within 65 km of one another were noted by Ryan and Wilczynski.

More detailed studies reveal call variation among populations, some considerably closer although females appeared not to discriminate among calls from different populations. Our work reveals that Bufo americanus calls may vary substantially on a microgeographic scale, over distances of 1 km or less, even in extremely similar environments. As we described earlier, males' calls resemble one another as a function of their genetic similarity.

Small, genetically homogeneous populations are most at risk of extinction due to a variety of causes. We suggest that conservation biologists might initially identify threatened populations by assaying variation in heritable behavioural traits. For each target species, baseline data are first needed to establish that behavioural differences are correlated with genetic differences, which may be determined either though pedigree analysis or genetic fingerprinting. With this information available, audio surveys might be used, not just for censusing, but also for monitoring genetic variation within populations.

Assessment of genetic variation by analysis of calls or other behavioural cues can potentially provide a quick means of identifying populations in possible danger without disrupting habitat or manipulating subjects. Where problems are suspected, genetic variation then can be assayed more precisely by molecular techniques to establish appropriate management plans.

Many traits, both morphological and behavioural, show evidence of fluctuating asymmetry when individuals are stressed. Environmental stresses, including pesticides, pollutants, acidification, and abnormal temperature regimes, appear to interfere with the developmental pathways that normally produce bilaterally symmetric traits. Populations with reduced genetic variation, highly inbred individuals, and individuals with low levels of heterozygosity are most likely to show evidence of these asymmetries.

As we have argued here, fragmentation depletes genetic variation and decreases the ability of individuals to cope with environmental perturbations.

Not surprisingly, individuals in fragmented populations show higher levels of fluctuating asymmetry than do those in undisturbed habitat. The measurement of fluctuating asymmetries thus constitutes another simple means of identifying populations that may be declining found that museum specimens of two species of tropical wetland frogs, *Litoria genimaculata* and *Litoria nannotis*, showed increasing evidence of fluctuating asymmetry in a variety of morphological characters during the months preceding the onset of their population declines.

We suggest that the monitoring of behaviours that are likely to show evidence of fluctuating asymmetry, such as locomotory abilities of tadpoles or adults, may even more readily identify populations that are being adversely affected by environmental and genetic stresses.

ASSESSING GENETIC VARIATION

Recent advances in genetic fingerprinting technology make direct molecular assessment of population variation feasible, inexpensive, and use relatively noninvasive procedures. DNA can be extracted from blood or from toe clips, which typically are obtained when marking amphibians in field studies. Should concern arise about possible effects of toeclipping dead skin shed by individuals can be used for genetic typing.

By use of the polymerase chain reaction (PCR), microgram quantities of DNA extracted from such tissues can be amplified repeatedly to generate sufficient material for genetic analyses. Thus, individuals no longer need to be sacrificed to genetically characterize populations. PCR methodology facilitates ongoing genetic monitoring of threatened species. Tissue samples can be stored in the field in ethanol or in other preservatives thus making genetic sampling possible even in remote locations.

As seen in our work on Bufo, a variety of approaches can be taken to assess levels of genetic variation. Analyses of variation in mitochondrial or nuclear genes may be more appropriate depending on the question of interest. Mitochondrial (mtDNA) genomes are small and thus more tractable, but they are also matrilineally inherited; thus they fail to reveal information on male dispersal and reproductive success. Analyses of mtDNA traditionally have focused on geographic population structure rather than on localized population variation.

Our work suggests, however, that amphibian mitochondrial genomes are sufficiently variable to generate maternal fingerprints that may be useful for studying genetic relationships within populations and fine-scaled population substructuring. Using PCR, primers can be designed to amplify the control region (D-loop segment), which typically shows the greatest variation.

Amplified products can be digested and characterized by fragment size to examine polymorphisms (RFLPs; restriction fragment length polymorphisms), sequenced to determine nucleotide substitutions or screened for variants by examining fragment conformation or stability (heteroduplexes, SSCP, DGGE, TGGE). Screening techniques can be used to design primers for specific mitochondrial and nuclear genes, such as those of the MHC, which may serve as especially useful markers in captive breeding. Nuclear DNA is more commonly fingerprinted for population studies. Currently several methods are available for analysing nuclear variation, each with advantages and disadvantages.

Until recently, variation was usually assayed by using probes that hybridize with particular sequences (VNTRs; sequences that consist of a variable number of tandem repeat nucleotides). Multilocus probes, such as those isolated by Jeffreys nonspecifically detect minisatellite loci containing these VNTR regions, which usually are homologous across taxa. This methodology can be used even to characterize individual tadpoles from pieces of their tails but larger quantities of tissue typically are required to type individuals confidently. Single-locus probers provide less ambiguous genetic data, which facilitate the computation of population parameters. The identification of even shorter repeat sequences, termed "microsatellite" loci, makes possible precise and meaningful estimates of genetic variation for conservation studies. Although inbreeding, population substructuring, and population size may affect interpretations the utility of microsatellites for population studies is undisputed. Microsatellite sequences can be amplified by PCR, so only minute tissue samples are required. Unfortunately, primers must be designed for each species (or species group) and for each locus, and the process is laborious and time consuming.

The use of random primers (RAPDs; random amplified polymorphic DNA) overcomes these difficulties, but suffers from problems of interpretation and repeatability. Nonetheless, RAPD primers quickly and inexpensively generate numerous genetic markers which can provide a useful measure of genetic variation that should allow conservation biologists to readily target populations of special concern. Kimberling *et al.*'s study of gene flow among *Rana pipiens* population serves as a good example of the versatility of RAPDs. With sufficient sample sizes, population parameters (degree of inbreeding and population subdivision, levels of heterozygosity, effective population sizes, and gene flow) can be estimated using these markers.

Differentiation can be partitioned within and among populations by using AMOVA (analysis of molecular variance) procedures originally derived for analyses of mtDNA haplotypes, but equally applicable to RAPD data. Recent work raises the possibility that the ease of use of RAPDs may be combined with the precision of microsatellites by using anchored primers to amplify short, interspersed repeats common to the genomes of most species.

CONSERVATION STRATEGIES

We suggest that surveys of amphibians routinely should include genetic monitoring Programmes using noninvasive methods such as those we described here. Levels of genetic differentiation among populations need to be determined to identify appropriate management units and to determine source populations towards which conservation efforts should be directed.

Populations characterized by low levels of genetic variation merit special attention because of their increased vulnerability to environmental perturbations and pathogens. Behavioural bioassays and analyses of fluctuating asymmetry may facilitate the identification of populations contaminated by toxicants or exposed to other stresses. Searches for particular pathogens in contrast, may be less fruitful because amphibians in declining populations appear to succumb to organisms normally present in their environment. Managers potentially can boost levels of genetic variation within populations by facilitating gene flow between populations. This can be done either by providing corridors or by conducting periodic translocations among separated populations.

As few as 0.5 migrants per generation among populations should reduce levels of inbreeding and eventually restore genetic heterogeneity within populations. In practice, attempts to translocate amphibian populations often have not been successful and translocated populations rapidly lose allelic diversity. The translocation of individuals to maximize genetic diversity within populations may represent a better solution. To prevent the transmission of pathogens, individuals must be thoroughly screened before relocation.

Outbreeding depression represents another risk, especially if transferred animals are adapted to different environments. Clearly, translocations should not be conducted between populations characterized by genetic differences that may qualify them as separate management units. Despite the potential pitfalls, carefully conducted translocations offer the promise of reversing the loss of genetic diversity that contributes to amphibian declines. Reports of population declines of amphibians, especially frogs and toads, have increased in recent years. Population extinctions may be now occurring with increasing frequency. Amphibian populations are adversely affected by the destruction or fragmentation of habitat, ultraviolet radiation, changing weather patterns, acidification, herbicides and pesticides, pollutants, and the introduction of exotic predators or competitors. Yet these factors do not explain all population declines, and many cannot be accounted for by these factors.

Environmental changes that appear harmless nonetheless may cause sublethal stress. Multiple factors may act synergistically to impair immune function and increase susceptibility to disease. Disease is associated with declines of many amphibian populations.

Because genetically similar individuals share immunological vulnerabilities, pathogens can more readily build up to critical epidemic levels when populations are genetically homogeneous. Many declines have occurred in regions where amphibian populations are fragmented. Fragmentation disrupts metapopulations, increases the probability that local populations will go extinct, and decreases the probability of recolonization. Recolonized populations will suffer reduced genetic variation. Amphibians are highly philopatric, which further reduces gene flow between populations. Levels of genetic homogeneity within populations may be high, leading to inbreeding depression, especially in stressful environments. Despite the potential for inbreeding, our field studies on toads demonstrate significant inbreeding avoidance in natural populations.

Closely related mates have similar advertisement calls, and call similarity between males is correlated with their genetic similarity. By means of playback experiments, we demonstrated that females can discriminate between calls of males more and less closely related to themselves, preferentially choosing the more distantly related male. Kin recognition in some taxa is mediated by cues encoded by histocompatibility alleles.

Variation in these cues may result from selection for disease resistance. Behavioural mechanisms that promote outbreeding may offset, at least partially, the depletion of genetic variation expected in small populations and can boost disease resistance.

11

Environments with and without Large Carnivores

During the earlier part of this century when conservation efforts focused on the prey of large carnivores, the loss of "top" carnivores was a concern to only a small minority of biologists. Although naturalists like John Muir and wildlife managers such as Aldo Leopold were disturbed about the rampant destruction of predators, the prevailing view, particularly in North America, was that predators destroyed prey valued by humans. In 1914, for instance, some 40 years after the establishment of the world's first national park, Yellowstone, the acting superintendent granted permission "to shoot trap, or poison mountain lions, wolves, and coyotes in the Park great care will be used not to shoot so as to injure anything except the animals mentioned".

Not only have campaigns to decimate carnivores been commonplace in the past, but many continue today. Thus, these conservation efforts can hardly be considered very active; it seems more reasonable to say that past conservation focused on the prey of carnivores, and the removal of carnivores was largely ignored. Fortunately, a different ethic is emerging now that more fundamental relationships between predators and prey are understood.

Environments with and without large predators have been used to assess direct impacts of predation and/or the loss of such processes. Where, for instance, herbivores have been released from predation, trophic cascades may occur, as seen in northern Pacific Ocean kelp forests which were severely impacted by the grazing of marine invertebrates after humaninduced localized extinctions of sea otters *Enhydra lutris*. In boreal and arctic systems, wolves and grizzly bears may depress prey populations, particularly moose, although environmental factors such as snow or cold interact with prey density to affect population size. As a result of predator-prey studies, it has been possible to place limitations on human harvests so that local prey populations are not depleted by hunting.

Recent research on prey abundance and population dynamics in the presence and absence of predators also suggests how wildlife management can contribute directly to local economies in developing countries through ecotourism and direct harvest and to the welfare of indigenous people.

Understandably, conservation involves more than science alone, and public perception plays a prominent role in how some components of biodiversity are managed. This is particularly evident in North America and Europe, where species reintroductions have occurred and mammalian carnivores have a disproportionate potential to impact public opinion most strongly.

EFFECTS OF LARGE CARNIVORES ON PREY BIOLOGY

Predation has ecological, evolutionary, demographic, and behavioural effects on prey populations. Where large carnivores are present, prey often occur at low densities, an ecological situation referred to as the "predator pit"; this situation is supported empirically by data on wolves and moose. In lightly exploited North American boreal and arctic systems, wolves depress prey densities, but where wolves have been exterminated or are heavily exploited, moose can be up to three or four times more abundant. Changes in predation pressure may also affect prey population density indirectly through compression of prey into safe zones due to human predation, or by giving them fewer competitors.

In evolutionary terms, guppy life-history parameters have changed genetically in less than 15 generations with alterations in predation pressure. Among other taxa, predation regime has altered sexual advertisements, mate guarding, synchronous swarming, and habitat use. In environments where large predators have been eradicated, changes in prey demography can be pronounced. For instance, where predation has been relaxed, individuals which might otherwise die experience higher survival and even physically deformed individuals may reproduce.

Finally, study of prey in small reserves lacking natural predation has facilitated the study of several paradigms in behavioural ecology. In both red deer *Cervus elaphus* and bison *Bison bison,* individuals have been followed throughout their lifetimes on the Scottish Island of Rhum and on American grassland parks in Montana and South Dakota, respectively, and from these studies, topics like dominance, reproductive synchrony, and parental investment have led to the refinement of evolutionary theory. But the elimination of large carnivores has also removed the possibility of understanding ecological complexities that may affect social processes; for smaller species such as pronghorn *Antilocapra americana,* it is possible to consider predation as a selective force because medium-sized predators still occur but for larger prey species, this is not always the case.

Bison on the National Bison Range (NBR), a protected reserve in Montana, highlight the dilemma. There they have no natural predators and are guaranteed relatively abundant food because they are rotated among paddocks. Excess bison are rounded up and sold annually, which results in relatively low density, and a fecundity rate (calves/cow) of 80-90 per cent, which exceeds that in less managed South Dakotan populations in Wind Cave (WCNP) and Badlands (BNP) national parks.

The NBR population is also characterized by more synchronous births, with 50 per cent occurring with 13 days; at WCNP, the same 50 per cent point occurs in 21 days and at BNP it requires from 20 to 27 days. Depending on the density at a given site, births can be interpreted as being synchronous or asynchronous. If loss of predation affects behavioural phenomena, then part(s) of the repertoires of wild animals may be changed or lost even before being described. For instance, animals are remarkably deft in discriminating among dangerous sounds and smells distinguishing visual from olfactory cues and differentiating aerial from terrestrial predators. These behaviours might disappear in the absence of predation.

EFFECTS OF LARGE CARNIVORES ON PREY BEHAVIOUR

Prey might respond to different levels of predation pressure in one of several principal ways. Responsiveness in this case is any measurable behavioural change and can vary from that associated with strictly behavioural events such as recognition of predators, transmission of cultural patterns of predator avoidance, or active defence to those more tightly linked to ecological conditions such as group formation and group size, vigilance, or shifts in habitat use. In scenario A predation pressure is constant, though it may operate at levels varying from high to low.

For simplicity, I assume that predation has been intense, as characterized by specialists like black-footed ferrets *Mustela nigripes* on prairie dogs *Cynomys ludovicianus*, cheetahs on Thomson's gazelles *Gazella thomsoni*, and grizzly bears on moose calves. Scenario B depicts the truncation (extirpation) of effective predation, a situation found in many national parks or protected areas where predators such as lions *Panthera leo*, spotted hyenas *Crocuta crocuta*, or wild dogs are locally extinct and human hunting is prohibited.

Examples include comparatively small reserves like Matopos in Zimbabwe or Waterberg Plateau in Namibia, or places where wolves and grizzly bears are extinct such as in Yosemite in California and Big Bend in Texas. Scenario C involves the recent restoration of large carnivores (the time frame remains unspecified). The restitution of predation as a selective force can be of two primary types: predators to which the prey have been exposed in the past, or those which are totally novel (*i.e.*, no evolutionary history of sympatry). Wolves reinvading Russia's Voronezh Reserve after an absence of nearly two decades represent the former. Whether prey have immediate (proximate) memory or some form of innate recognition of their historic predators is uncertain, but responses might be very different from those of prey encountering a new predator for the first time. Komodo dragons *Varanus komodoensis*, which may have once preyed on extinct dwarf elephants, now kill rusa deer *Cervus timorensis*, which did not evolve with this reptilian carnivore and may show inappropriate antipredator responses.

Both pumas *Felis concolor* and coyotes have expanded their ranges in areas of the Great Basin Desert and southeastern Quebec during the last 100 years, where local populations of porcupines *Erethizon dorsatum*, bighorn sheep *Ovis canadensis*, and caribou *Rangifer tarandus* may be ineffective at dealing with them. While such range expansions by mammalian carnivores are due to recent anthropogenic events, the exposure of naive prey to novel suites of predators is common throughout history.

Human colonization of the Americas, the spread of large felids (*Panthera* spp.) across Beringea into the Yukon and Alaska, and/or the replacement of thylacines by dingoes after introduction into Australia by humans some 4000 years ago all represent cases in which prey were forced, often unsuccessfully, to respond to a predation pressure that presumably differed from that of their ancestors. Among the methods for detecting behavioural changes associated with the loss of or change in predation, the comparative method is one of the best. For instance, in areas where predation has been relaxed, ground squirrels and salamanders respond differently than do their predator-savvy conspecifics.

Contrasts between island populations of macaques with and without tigers and leopards *Panthera pardus* has led to the suggestion that macaque group sizes have changed where predators have been lost. Yet it is difficult to exclude the role that other factors, notably food quality or distribution, may have played in influencing group size. Another difficulty arises when predators are extinct but exerted strong selection in the past. The loss of eagles for about 4000 years from Madagascar's contemporary fauna is a case in point. Two lemurs, ring-tailed *Lemur catta* and Verreaux's sifaka *Propithecus verreauxi*, show strong antipredator responses to birds, behaviour that has been interpreted as the result of past selection.

But without excluding other possibilities such as contemporary predation by other birds, it is not possible to say much about predation-related effects. Another possibility for examining how predation affects behaviour is to contrast responses in closely related species, as has been done for endemic flightless birds in New Zealand in order to determine responsiveness to introduced mammalian predators. In short, although components of mammalian prey behaviour have been linked to the relaxation of predation as a selective force, appropriate controls are usually absent. Empirically grounded field experiments are therefore the best means of understanding how prey respond to both the absence and presence of predators.

EXPERIMENTAL DATA

As a result of the widespread extermination of large terrestrial carnivores on all continents, opportunities exist to examine issues in prey behaviour where predation has been relaxed. Moose, for example, have a circumpolar distribution in environments that vary from predator-rich to predator-depauperate. In the Sikhote-Alin Mountains of the Russian Far East, moose live in a three-predator system that includes grizzly bears, wolves, and Siberian tigers and in Alaska they coexist with two predators (grizzly bears and wolves). At sites in the Rocky Mountains, moose live where these same two native predators have been eliminated, exist, or have been replaced by subsistence or trophy hunters.

Predation on calves has been intense in Alaska's Talkeetna Mountains, with about 67 per cent dying before 2 months of age (W. Testa, personal communication), but calf survival exceeds 95 per cent at two sites in the southern greater Yellowstone ecosystem, one with human hunting of adults only and the other in the absence of predation because grizzly bears and wolves are lacking. Assuming predation intensity has been relaxed where large carnivores have been eliminated, to what extent might moose behaviour have changed? If predation has shaped the ability of prey to recognize predators then in its absence naive prey should fail to respond to cues suggestive of their major predators. If unfalsified, the hypothesis suggests that the immediate behaviour of naive prey has changed in the absence of predation. Should the antipredator behaviours of savvy and naive prey (*i.e.*, scenarios A and C) remain similar, the responses may either be obligatory or prey have had insufficient time to respond to selection.

Among the ways to detect prey responsiveness to potential predators are observations of direct encounters and through field playback experiments, since the chances of witnessing direct interactions between large predators and their prey are possible in only a few areas of the world. More indirect procedures might include assessing prey behaviours associated with predator detection in naturally varying environments while controlling possible confounding variables. Thus, to examine potential behavioural changes associated with differing predation regimes, I have relied on the behaviour of nonhunted female moose after presenting auditory and olfactory signals of wolves, grizzly bears, and other species.

The experiments were conducted by playing different sounds from a speaker placed 50-200 m from moose. The olfactory trials were conducted by soaking snowballs in urine or wrapping them with seat in biodegradable toilet paper or netting and then tossing them to within 5-10 m of prey subjects. Both types of field trials were conducted on windless days, when ambient temperatures varied from -2° to 5°C and when animals were feeding. The response variable was the proportion of time that animals fed per 180 s.

Preliminary results are tantalizing, for they suggest that in the absence of active predation for about 50-75 years, female moose in Grand Teton National Park show only weak responses to the sounds and odours of wolves. This is not the case in Alaska, where wolf predation is still present. Systematic playbacks of both wolf howls and exposure to wolf urine depressed feeding rates of the predator-savvy Alaskan females more than it did the predator-naive Teton females. Similarly, raven vocalizations depressed feeding rates more in predator-savvy than in predator-naive moose.

This is intriguing because predator-savvy prey might be expected to associate ravens with the presence of carcasses, perhaps indicating that wolves or other large carnivores are nearby. But where raven densities are the highest recorded such as in the southern Greater Yellowstone region of my study area and large carnivores are absent, moose failed to show feeding reductions.Exposure to grizzly bear feces, irrespective of the past associations of female moose with carnivores, depressed feeding rates similarly, and more so than exposure to wolf urine.

But there were no such reductions in Teton moose when exposed to wolf urine. If moose respond more intensively to one type of predator than another despite 50-75 years of no exposure to either, the possibility of innate recognition may persist. It may also be that one type of predator in this case grizzly bears, has had a stronger evolutionary influence than others. In Alaska, grizzly bears continue to prey more heavily on moose calves than do wolves.

While preliminary, these results have implications for understanding how prey recognize potential threats from predators. White-tailed deer *Odocoileus virginianus* congregate along wolf territorial boundaries and chital *Axis axis*, moose, and vervet monkeys *Cercopithecus aethiops* all decrease predation risk by associating near human dwellings where tigers, leopards, and wolves are less likely to occur. If effective predators are lost from these and other systems, the clustering of individuals or sexes to decrease predation risk, patterns of spatial association, and habitat use may all change.

STRENGTHS AND WEAKNESSES OF A BEHAVIOURAL ECOLOGICAL APPROACH

As a construct for understanding immediate issues in

conservation, large carnivores offer advantages and disadvantages. On the positive side, they elicit extraordinary attention from the lay public and politicians, are economically and ecologically valuable and, as an educational tool, they are usually unsurpassed.

But, as subjects of scientific study, large carnivores are problematic because individuals range over large areas, populations are small, and experimental manipulations are difficult even where feasible. Pragmatically speaking, it is obvious that as human impacts increase globally, progressively fewer environments will maintain assemblages of large carnivores, and it would be surprising if the biology of their prey has not changed or is not changing.

If we want to understand our future ecosystems without large carnivores or develop serious plans to restore carnivores, we must anticipate behavioural and ecological relationships in prey by seizing opportunities for comparative approaches while study is still possible. Among the key issues that involve predator-prey dynamics once carnivores are either restored or lost are the functional responses of prey populations. With carnivores restored, their primary prey are expected to decline, whereas prey should increase if their major predators are eliminated.

Opportunities also exist to understand how prey respond to predation; it is here that approaches steeped in behavioural ecology can play a role. For instance, rufous-hare wallabies *Lagorchestes hirsutus* became extinct on the Australian mainland not because they were unable to cope with predation by native species, but because of introduced carnivores—red foxes and feral cats—with which they were unfamiliar. Knowledge of antipredator behaviour has been critical in designing plans to achieve viable wallaby populations; in this case, the effort has focused on teaching wallabies to avoid predators they have not evolved with, a situation also employed with other endangered mammals. In today's anthropogenically disturbed world, predation comes in two primary forms, those in which predators are either familiar or nonfamiliar to possible prey. The former, involving species in which predator and prey evolved together, was the typical situation, whereas the latter is likely to become increasingly common. Behavioural ecologists have generally focused on natural systems with native species because the discipline has traditionally been concerned with evolutionary aspects of behaviour.

What eventually may prove more central to conservation will be an increased focus on exotic predators—the countless cases of foxes, dogs, cats, mongooses, birds, snakes, and fish that have been disseminated by our global carelessness and which have devastating effects on local prey populations. An understanding of behavioural variation and responsiveness to introduced and native species would do much towards enhancing conservation.

For instance, populations of the California newt *Taricha torosa* apparently remain viable where exotics have not entered the system. Their larvae are able to avoid predation by native species but cannot avoid predation by introduced mosquitofish *Gambusia affinis* and crayfish *Procambarus clarkii*. Additionally, where the aim is to assure the persistence of local prey populations, it will be important to predict both behavioural and subsequent demographic responses to specific types of carnivores, irrespective of whether they are recolonizing or novel.

And, in cases where prey populations are reintroduced into former habitats, it may be possible to enhance their viability by knowing about predator avoidance capabilities. Numerous anecdotes have consistently been raised in my discussions with management agency personnel in which reintroductions involving species as diverse as eland *Taurotragus oryx* and porcupines were unsuccessful because individuals were predatornaive. Although anecdotes may differ from reality and be susceptible to memory failure over the years, the message is clear: more can be done about enhancing population viability by designing research that takes into account the social history of founding populations and the extent to which individuals may be familiar with past and present predators. I have argued that behavioural ecology has a role in the conservation biology of prey assemblages, but there are obvious constraints in developing appropriate study designs with large carnivores and large prey. Also, a lack of replication is likely to characterize most field studies, but this problem may be partially circumvented through the use of meta-analyses. And it is not totally clear how field experiments, such as the ones I described using the sounds and smells of novel and familiar carnivores, simulate real predator recognition, nor is it clear the extent to which behavioural responses are linked to demographic ones. If colonizing (native or exotic) or recolonizing carnivores encounter naive prey, the potential for local extinction exists. Indeed, evidence from porcupines and caribou suggests that native prey may be differentially susceptible to steep population reductions when encountering native but novel carnivores for the first time. Large macropods protected by fences from dingoes may also suffer disproportionate mortality relative to predator-savvy conspecifics when dingo reinvasion occurs, although little is known of how prior history affects individual kangaroo antipredator responses.

RECOMMENDATIONS

There are at least three tactics that biologists and managers of reserves should consider with respect to prey populations when planning for reintroduction of prey or the restoration, recolonization, first-time arrival, or loss of large carnivores.

PLAN FOR RESTORATION

Efforts should include anticipating natural recolonization and/or human-assisted efforts to restore previously extirpated species. If restoration efforts proceed, the effects of predation and when or if the system attains some equilibrium point need to be determined. Two endangered species offer real-world examples. Both golden lion tamarins *Leontopitheus rosalia* and black-footed ferrets *Mustelia nigripes* have been reintroduced at sites where populations were jeopardized or already extinct. Because individuals targeted for reintroduction lacked experience with predators in their new natural environments, a primary concern was whether these individuals would deal effectively with potential predators. Behavioural ecologists were instrumental in designing Programmes to reestablish antipredator behaviours.

NOVEL CARNIVORES

Where restoration of past communities is not feasible (due to agricultural or other economic interests) but colonization by different carnivores is a possibility, it will be important to anticipate responses of prey in order to maintain viable populations. Elaboration of two examples is illustrative. One hundred years ago, mule deer *Odocoileus hemionus* were rare, and pumas were presumed virtually nonexistent in the Great Basic Desert.

As domestic stock and feral species altered ecological communities through their effects on vegetation, deer populations erupted, followed by pumas. A variety of prey are now consumed, including some which may be highly susceptible to predation, such as bighorn sheep. Despite quills for defence, porcupines were driven to near local extinction in one population. Thus, while still a native species, the range of pumas has expanded in part due to the indirect effects of pastoralists, which caused habitats to change.

Carnivore range expansion has occurred elsewhere as well. Expansion of coyote ranges coincides with the extinction or reduction of wolves. In southeastern Quebec, the viability of remnant caribou is jeopardized, perhaps because coyotes are effective predators of juvenile caribou. If further study substantiates this, knowledge about caribou antipredator responses could lead to important management considerations. For instance, where prey do not possess the ability to avoid novel predators and the viability of the former assumes precedence, control of predators should be an option.

CARNIVORES WITH HUMANS

Where restoration has not been possible and effective predation no longer occurs, pragmatic conservation biologists will have to deal with a politically sensitive issue: how to balance the abundance of single species with ecological diversity. If the latter is important, then population control will be critical whether by replacing mammalian carnivores with human predation or by contraception or translocation. With regulation by humans, antipredator behaviour in the remaining population may change.

From a purely conservation perspective, a critical ecological issue concerns the extent to which the replacement of large carnivores by humans alters other components of the community. In the absence of some regulation by either natural predators or humans, populations even in large areas like the greater Yellowstone ecosystem may overshoot their food ceiling. At some point, decision makers will be forced to make difficult choices between protecting the primary prey of large (and currently extinct) carnivores or protecting that ecosystem's biodiversity. For instance, abundant herbivores damage the vegetation of riparian zones which in arid environments may harbor most of the region's diversity, and as a consequence both avian and lepidopteran species may decline.

If the goal of managers, then, is to understand the spatial dynamics of prey or causes of juvenile mortality, to predict susceptibility to new or past predators, or to restore prey populations, then approaches employing behavioural ecology will yield insights into how systems operate. By understanding decisions that individuals make about foraging, socializing, and reproducing, it will be possible to understand population-level responses to predation. The earth is likely to lose large carnivores in the near future, a result that inevitably will alter pressures on prey.

In the United States and Mexico alone, large and economically important species such as grizzly bears and wolves have experienced a 99 per cent reduction in their historic range. Because predation is a major selective force and affects ecological dynamics, the altered environments in which we now work are either missing dominant selective forces or have been replaced by humans. The lack of natural carnivores at the top of food webs is likely to have repercussions on behavioural phenomena of prey. These may include reproductive synchrony and grouping as antipredator strategies, altered spatial distribution, and altered responsiveness to potential predators. Comparative data on the behaviour of female moose at a predator-rich site in Alaska and a predator-depauperate site in the Rocky Mountains imply that where predation has been relaxed, prey reduce their reliance on auditory and olfactory cues associated with predators that were extirpated only 50-75 years ago. Although preliminary, such data suggest that prey may have to learn how to avoid some predators.

If the loss of some large carnivores alters selection pressures, then so must the addition, range expansion, and restoration of others. Expanding populations of native species such as pumas and coyotes and the introduction of exotics such as dogs, cats, foxes, and mongooses have disproportionate effects on prey that have evolved in their absence.

12

Behavioural Ecology of Hunters

SUSTAINABLE SUBSISTENCE HUNTING

Although studies have demonstrated that a species' intrinsic rate of increase, resulting from differences in body size and phylogeny, is a key factor influencing susceptibility to overharvesting, it is clear that other factors are also involved. The rate at which a species is harvested will depend on a range of behavioural factors both on the part of the prey and the hunter. Behavioural ecology may therefore help in the development of sustainable harvesting management plans by determining which behaviour patterns contribute to inter- and intraspecific variation in vulnerability to harvesting. It may also help us determine the consequences of harvesting, in terms of its effects on the social organization and behaviour of prey, and thus predict how particular prey species will respond to different levels of harvesting. Appreciating the factors influencing human hunting behaviour may enable managers to understand what determines prey selectivity and how variations in habitat, prey availability, and hunting methods influence the impact of hunting on prey populations. In the next section, some behaviour patterns that are likely to influence either inter- or intraspecific vulnerability to harvesting are described. In the following section, the determinants of human hunting behaviour are explored, and the consequences for prey populations are outlined.

SOCIAL BEHAVIOUR

In many prey species, grouping has evolved as a form of

predator defence. The antipredator advantages of grouping include improved predator detection and defence as well as a reduced risk of predation as a result of the dilution effect. The dilution benefits of grouping rely on the fact that the number of attacks and the number of individuals killed in a predator-prey encounter does not increase in proportion to group size. Overall, the benefits of grouping primarily accrue to species inhabiting open habits, as the difficulties of maintaining groups and monitoring other group members are considerable in forested habitats.

In addition, individuals living in forested habitats often rely on concealment, which is easier when alone or in small groups. Few data are available on the effects of grouping on vulnerability to human predation, but a number of social species make use of group defences to reduce vulnerability to human hunters (*e.g.*, musk oxen *Ovibos moschatus*), and some authors report hunters being reluctant to harvest grouped prey because of the risk of injury (*e.g.*, the Achuara avoid large groups of white-lipped peccary *Tayassu pecari*).

The increased alertness of large groups of prey is likely to result in hunters favouring individuals on their own or in smaller groups, just as natural predators frequently do (*e.g.*, cheetahs *Acinonyx jubatus* hunting Thomson's gazelles *Gazella thomson*). There is some evidence that group living species may be more vulnerable to human hunters under certain conditions. Groups are easier to locate than solitary individuals, and once the group has been found, more than one individual can be killed. For example, describes how the herding behaviour of buffalo *Syncerus caffer* is exploited by hunters to eradicate local herds in cooperative drives into snares lines.

Similar techniques were used by Chipewyan to catch large numbers of caribou *Rangifer tarandus* in Alaska. Hunters may also make use of the social behaviour of prey species to attract them. For example, M. Alvard (personal communication) reports that if Piro hunters in Peru catch a social primate that is still alive, they will try to force it to make distress calls that attract other members of the troop. Once these other individuals have been lured into range, they can then be killed. A similar technique is used to catch capybaras *Hydrochaeris hydrochaeris*; one hunter carries a young capybara as a lure while another walks ahead with a torch and a club to intercept any capybaras that approach. Differences in grouping behaviour between age and sex classes may result in differential predation rates. For example, the preference of hunters to stalk solitary individuals is likely to result in increased predation of males, which are more likely to be on their own or in small groups.

Adult blue duikers *Cephalophus monticola* usually live in family groups consisting of one male, one female, and their offspring younger than a year and a half, while subadults live singly and tend to be found away from preferred habitats. Consequently, Hudson argues that because net hunters tend to set nets in areas with high duiker densities, subadults are less likely to be caught. Nevertheless, a repeat netting session in a recently harvested area is likely to catch a high proportion of subadults which have moved in to occupy the vacant territories.

FLIGHT BEHAVIOUR

Variation in vigilance levels and flight distances is likely to predispose different species and different age or sex classes to selection by hunters. Females of sexually dimorphic or polygynous species are generally more vigilant and likely to flee from predators than are males probably as a consequence of the presence of young, relatively smaller body size, and/or group size differences. Such differences are likely to be manifest when humans are the predators.

For example, male Thomson's gazelles flee at shorter distances from humans than females (C. FitzGibbon, personal observation), whereas female black rhinoceros *Diceros bicornis* flee farther from approaching humans than males and consequently may be more prone to human predation. Males may also be more vulnerable during rutting because of reduced vigilance towards predators and increased focus on defending access to females (*e.g.*, in moose *Alces alces*).

Similarly, archaeological research on Steller's sea lions *Eumetopias jubatus* suggests that prehistoric hunters primarily killed males (75 per cent of the harvest), probably because they were hunted during the breeding season when males maintain and vigourously defend territories.

Females and juveniles flee into the ocean when attacked, but males stay and defend the beach. The inability to flee as fast as other individuals may predispose pregnant females to harvesting, particularly when hunting techniques rely on active pursuit. Variation in flight distances may also influence the vulnerability of different age classes. Behavioural observations by Dubost suggested that young blue duikers less than 10 months old were more likely to remain quiet and stay hidden rather than to jump and run when startled. This makes them less vulnerable to net hunters such as the Aka.Bodmer suggests that older peccaries may be more wary of hunters and therefore less likely to be caught, although his data did not confirm the suggestion. Archaeological research suggests that nursery groups of newborn harbor seals *Phoca vitulina* were exploited to the near-exclusion of adults of this species because the adults tended to be more vigilant than the naive young.

TEMPORAL ACTIVITY PATTERNS

Although it is predicted that nocturnal species are less likely to be caught by hunters than diurnal ones simply as a result of reduced opportunities for capture, few quantitative data are available to confirm this. Nevertheless, nocturnal species, particularly primates, make up a small proportion of captures in most locations; for example, Britton reports that sloths are rarely harvested because of their nocturnal behaviour patterns. Alternatively, hunters may make use of the fact that animals can be blinded at night and use spot lights to facilitate capture. For example, hunters in Arabuko-Sokoke use strong flashlights to blind duikers temporarily. In addition, species with regular behavioural routines, such as visiting waterholes at particular times, may be more susceptible to hunters.

TRAIL AND BURROW USE

Species that make trails through vegetation or use burrows, particularly simple, single-exit ones, are more vulnerable to snaring than those species that have less-defined movement patterns. For example, one of the reasons four-toed elephant shrews *Petrodomus tetradactylus* are caught much more frequently than the golden-rumped species *Rhynchocyon chrysopygus* in Arabuko-Sokoke Forest, Kenya, is that the former make trails through the leaf litter along which snares can be set. Pigs are also vulnerable because they travel along trails. The sedentary and localized beaver *Castor canadensis* is susceptible to overtrapping because it can be easily snared at the entrance of its lodge.

ATTRACTION TO BAITS

Species that can be attracted into traps using baits, calls, or decoys are particularly vulnerable to harvesting. Baiting of primates and rodents is widespread, but it also applies to other groups. For example, hunters in Korup, Cameroon, attract bay duikers *Cephalophus dorsalis* by calling. Many American peoples focus on species that are attracted to their gardens—for example, the Mayas Afroecuadorians, and Cachi Ungulates are easily attracted to burnt grasslands, even a few hours after burning, and many indigenous people make use of this fact. North American Indians burn reeds along lakeside margins to increase root growth, attracting muskrats *Ondata zibethicus*. Hunters may also make use of animals' attraction to natural features such as water holes and salt licks. For example, Amazonian Indians wait in ambush for *tapirs Tapirus terrestris* attracted to such features.

RANGING BEHAVIOUR AND HABITAT CHOICE

Dispersal and ranging behaviour will also affect a species' vulnerability to harvesting, influencing the extent to which individuals move from protected areas into hunting zones and their ability to recolonize depleted areas. In predator-prey theory, dispersal of prey from predator-free patches (sources) into harvested areas (sinks) is considered a key factor in prey persistence. Harvested populations frequently survive only as a result of immigration from other areas. Species may depend on the preservation of refuges large enough to maintain a sufficient flow of recruits. Obviously, the dispersal ability of the prey species will influence the scale of patchiness required to maintain a population. Migratory species may be particularly vulnerable to overharvesting. This is partly because species that have regular movement patterns may be favoured as prey, given the ease with which hunters can locate and intercept them.

For example, the spring and fall migrations of caribou make them vulnerable to capture as they move through mountain passes. In addition, it is difficult to assess declines in migratory species, particularly when such species form large aggregations during migration. As a result there is some suggestion that even those societies that do practice some conservation ethic will overharvest migratory species, tending to take as many prey as possible when they do appear, knowing they may vanish tomorrow and not reappear for many months. The "Pleistocene overkill" sites in North America involved herding animals that were probably migratory. Even local, predictable movements can increase a species' vulnerability to harvesting. Large herds of caribou tend to stay close to lake shores and avoid areas without thick bush or woodland. In the summer, the caribou are plagued by warble flies and select higher areas that catch the wind. Any predictable movement increases the chances of hunters successfully locating and stalking them. In contrast, in winter, moose are more difficult to locate because they become confined to small feeding patches where the snow is thin, and they do not leave long trails.

Differences in ranging behaviour between males and females may result in intraspecific variation in vulnerability. For example, in Serengeti National Park, the majority of wildebeest *Connochaetes taurinus* caught in snares are males (88-98 per cent) because males tend to be at the front of herds as they move into new areas and consequently meet snares first. Males may in general be more vulnerable than females as a result of increased activity and movement associated with searching for mates.

RESPONSE TO HARVESTING

The way in which species alter their behaviour in response to harvesting can have a major impact on their vulnerability. Many species become more wary of humans, increasing vigilance levels and flight distances and altering movement patterns, which reduces harvesting rates. For example, geese quickly learn to avoid hunting areas in preference for protected sites capybaras become more nocturnal and increase their vigilance and flight distance and large Neotropical primates become extremely shy of humans whenever they have been hunted in the past.

In addition, where hunting pressure is intense, the capybara, a savannah-living animal, is found more frequently in forest. However, changes in behaviour may also have damaging effects on food intake and reproductive rates, with individuals trading off the benefits of reduced predation risk against the costs of reduced foraging time or access to high-quality food resources. For example, Paraguayan caimans *Caiman yacare* respond to harvesting by reduced egg guarding, which increases predation rates on eggs and reduces reproductive success.

Hunting may also break up family groups, resulting in reduced breeding success. For example, orphaned juvenile geese survive less well, and breeding is frequently disrupted by the loss of a mate as a result of shooting. In addition, successful breeding in social species may be group-size dependent, and thus removal of a few individuals from a group may prevent the group from successfully breeding again.

Also, solitary living, monogamously mated species that live at low density may be vulnerable to harvesting because of the time required to make new pair bonds if one member of the pair is killed. Yet, in other species, missing partners are replaced very quickly (*e.g., dik-dik Madoqua kirkii*), and the potential for disruption of birth rates is low.

IMPACT OF HUMAN HUNTING BEHAVIOUR

Understanding and predicting the impact of human

hunting behaviour on prey has centered around the use of optimal foraging models. Although these models were originally developed for understanding the factors influencing nonhuman foraging behaviour, their utility to researchers of human foraging behaviour rapidly became apparent. The underlying assumption is that either the nutritional status of an individual or the time needed to acquire necessary nutritional requirements determines fitness.

Models have been primarily used first to predict prey choice (*i.e.*, predict the food items the forager will attempt to exploit and those it will ignore in favour of continued search for more preferred foods), and second, when resources are clumped together in patches, to determine whether a forager will enter a patch and how long it will stay in that patch before moving on to another.

These optimal foraging models have been combined with population biology models to simulate the dynamic relationships between hunter-gatherers and their prey resources. These models can be used to predict how the abundance of hunters influences prey abundance and the conditions under which overexploitation is likely to occur. The model developed by Winterhalder *et al* however, suggests that neither the dynamic properties nor the equilibrium density of the foraging population can be predicted from only the biomass of prey and the requirements of the foragers.

Steady growth to equilibrium, damped cycles, stable limit cycles, and extinction of either the predator or its prey are all possible outcomes, but small changes in factors such as foraging effort and prey density can influence the dynamic properties of the system. Although the predictions of these models have yet to be thoroughly tested, it is clear that optimal foraging models provide a means to predict the impact of hunter-gatherers on prey populations under a variety of conditions and constraints.

EXTENT OF HUNTING VERSUS PLANT GATHERING/FARMING

A number of studies have noted that hunter-gatherers

devote large amounts of time and energy to hunting, even when gathering plant food or farming would have maximized energyreturn rates. For example, virtually all South American horticulturalists obtain much higher caloric return rates from farming than they do from hunting or fishing, yet most spend considerably more time hunting and fishing than farming. One obvious reason is that meat contains high-quality protein, the nine essential amino acids that the human body cannot synthesize, and fats that are important for the absorption, transportation, metabolism, and storage of fat-soluble vitamins.

Linear programming models, a method of solving for behavioural optima under specified contraints, have been used to predict the combination of food types that best satisfies the nutrient-maximizing or time-minimizing goals that best suit a particular society and have been used to understand why so much time and effort are put into hunting as compared to gathering. Gathered foods increase in the diet as hunted foods are reduced in abundance but hunter-gatherers will continue to hunt prey populations even when they become scarce.

As a result of the high demand for meat, humans are capable of having severe impacts on prey populations even when prey are scarce. The use of a population model built around optimal foraging models s1uggests that hunter-gatherers are likely to have a greater impact on prey populations when gathered foods are more abundant, allowing larger human populations to be maintained and the exploitation of hunted foods to remain high even when hunted foods are rare. Consequently, the model predicts that overharvesting is more likely in areas of high primary productivity. This prediction is counter to that of traditional overexploitation models, where prey extinction is more likely at low primary productivities. In addition, the model suggests that the key to the demise of hunted foods is not their productivity or abundance, but the productivity of gathered foods.

SELECTING PREY SPECIES

Prey-choice models predict that the prey type that yields

the highest rate upon encounter should always be pursued. Other lower-ranked resources should be included sequentially until the next-most-profitable resource yields a lower rate of return upon encounter than that which could be obtained by continuing to search for and pursue the more profitable items. Most of the tests of the models have been qualitative rather than quantitative, predicting directional tendencies in prey choice or diet breadth in relation to directional changes in parameters such as return rate or abundance. Predictions include the fact that large prey are generally more profitable than small prey (as long as handling costs do not increase in proportion to body weight), that low-ranked resources should drop out of the diet when search costs decrease and hence overall return rate increases and that diet breadth should increase with a decrease in the density of highly profitable prey items.

A number of studies have reported hunters' preference for large or medium-sized prey over smaller prey The preference for medium-sized prey is particularly marked in the case of hunters harvesting for commercial markets because carcasses often need to be transported over long distances and in addition are easier to sell and are economically more viable. This increasing preference for large prey as travel time increases is predicted by central-place foraging models, which are characterized by a pattern in which the forager makes an outward trip, forages, and then returns home with the captured prey. As radial travel time increases, these models predict that the optimal diet breadth is progressively restricted to prey species of value, and the optimal load for the return trip increases in size.

The body size of harvested prey is important in determining susceptibility to overharvesting because larger species tend to have lower reproductive rates and lower densities and therefore are more vulnerable to overharvesting. The only factor that may favour larger species is the fact that they tend to be distributed over a wider area. Consequently, local hunting may be less likely to exterminate a whole population than intensive harvesting of a smaller species.

The preference for larger prey depends partly on prey

abundance, with smaller prey becoming more profitable as large prey decrease in abundance. Consequently, Hamesand Vickers predicted that in zones of high hunting pressure, hunters should shoot large animals when they are found, but also prey on the more abundant smaller game, and in zones of low hunting pressure where large game are more abundant, hunters should be more likely to ignore small game.

Testing these predictions on the hunting strategies of three Amazonian societies showed that in all three cases, the proportion of large game to small game hunted decreased with increased hunting pressure. Among the Yekwana and Yanomano, significantly fewer kills of small game (pacas *Agouti paca* and armadillos *Dasypus novemcinctus*) occurred in distant zones than would be expected given the probabilities derived from binomial distribution theory. Maximizing mean daily meat returns over the long-term (long-term rate maximization) may not be the most suitable measure for determining an optimum foraging strategy because the regularity with which meat is obtained (*i.e.*, the short-term temporal variation) may be equally important.

Although specializing on larger game may maximize mean daily return rates, success rates for big game hunting, measured as the chance of acquiring a carcass on any particular day, are often lower than for smaller prey. Hunters can reduce the variance in meat supply by including smaller game in the suite of prey they exploit, or when large food packages are acquired by different hunters, by sharing carcasses between them.

Thus, sharing increases the emphasis on large prey items, whereas reducing the risk of failure on any day puts emphasis on smaller prey. This may be important when small children need to be fed or when the risk of starvation is high. The need to provide some meat reliably each day, if only a little, is one reason elephant shrews are such popular prey in Arabuko-Sokoke.

Despite individual preferences for larger bodied species, hunters often harvest more smaller species particularly those that reproduce rapidly, because these species tend to maintain high densities and therefore are encountered more often and so are more easily caught. For example, Bodmer studying hunters in Tahuayo, Peru, found that in terms of the actual numbers of animals taken, harvests were positively correlated with reproductive productivity. An examination of a number of studies demonstrates that this may also apply in other areas. Another reason hunters may be constrained from taking as many large prey as they would wish is that the harvesting of large prey often depends on cooperation between hunters. For example, while individual Aka hunters from central Africa regularly trap small rodents, setting nets for larger prey, such as duikers, requires the cooperation of more than 30 adults.

SELECTION OF SPECIFIC AGE CLASSES

Assuming that handling costs are equal across different age categories, optimal foraging theory predicts that adults are more profitable prey than juveniles because they are larger. In most subsistence harvesting situations, however, there are few opportunities for age-biased selection. Bodmer for example, states that low visibility in dense Amazonian forest results in animals being harvested randomly with respect to age.

Alvard found that harvests of peccaries (*Tayassu* spp.) and red brocket deer *Mazama americana* by Amazonian hunters were approximately random with respect to age (although the age classes were understandably broad). Marks found that Valley Bisa hunters concentrated on adult animals, as they provided better returns for their hunting effort. In contrast, capybara harvesting, which primarily takes place in open savannah habitat, mainly removes larger adults, which results in the age structure of harvesting populations being heavily biased towards younger individuals. Differences in age-class susceptibility may also affect selectivity.

Theoretical considerations suggest that the impact of harvesting on a population's reproduction rate can be minimized by harvesting animals with a high risk of mortality or low reproductive potential. Studies of the reproductive success of known aged individuals (*e.g.*, red deer *Cervus elaphus*) suggest that, in general, harvesting prime-aged individuals will have a greater effect on the population than harvesting either young or old animals. In social capybaras, younger females have relatively low rates of pregnancy and smaller litter sizes than older females so the selective harvest of larger, older females is likely to have a detrimental effect on reproductive rates.

However, behavioural ecological studies suggest that the effects of removing large proportions of subadult animals may have more severe consequences than theoretical studies suggest, particularly in social species.

Subadults often contribute significantly to the reproductive success of the breeding adults by looking out for predators, finding food, and defending the territory, and in some cases they may be essential for successful reproduction (*e.g.*, in social mongooses, a prey species in East Africa).

SELECTION OF DIFFERENT SEXES

Assuming that handling costs are equal for male and female prey, optimal foraging theory predicts that the larger sex (usually the male) is more profitable and therefore should be hunted most often.

As with age selection, in most subsistence harvesting situations, there are few opportunities for sex-biased selection. Trapping or netting is rarely sex specific, and even when shooting the prey, limited visibility often makes identification of males and females difficult.

There are few data available on the sex ratio of prey harvests from subsistence hunters. Marks found that Valley Bisa hunters in Zambia killed mainly male ungulates; Alvard found little evidence for preferential selection of male prey by the Piro in Amazonia; whereas Peres reported that hunters in western Brazilian Amazonia preferentially targeted breeding female primates because the infants they carry may be captured alive and sold as pets.

Irrespective of hunters' preferences for males and females, it is possible that one sex may be more vulnerable because of differences in behaviour or size. For example, there is some evidence that female primates are more vulnerable than males because of the mobility costs of pregnancy and carrying infants. Male-biased harvests have been shown to increase the

MSYs of populations both in theory and in practice under the assumption that many species have a polygynous social structure in which each male mates with several females, and consequently that a femalebiased population sex ratio will not have a detrimental effect on female reproductive rates. Some of the effects of female-biased populations may well be advantageous—for example, reduced aggression and mortality among males and reduced harassment of females by bachelor males.

Behavioural ecological data suggest that the effects of male-biased harvests may be more detrimental than originally believed, particularly for certain species. Potential deleterious effects include disruption of territorial and group structure, increased mortality of young born out of season, artificial selection for inferior males, or an inadequate number of males to inseminate females.

Moreira modeling the effects of harvesting on capybara populations, found that hunting had greater impacts on populations when it was sex biased (either way) than when it was not because of the different roles of males and females in this social species. In addition, although Fairall argued that in impala highly male-biased hunting is unlikely to limit female fecundity because there are "excess males in bachelor herds," Ginsberg and Milner-Gulland point out that detailed behavioural data on known individuals suggest that an excess of males in impala cannot be assumed from observations of bachelor groups.

In a single breeding season, a territory may be held by several males. Hence excess males may actually be breeding males who have lost or not yet gained a territory.

There is some evidence that low proportions of males in ungulate populations can result in reduced female fecundity. Removing a high proportion of adult males might result in breeding by partially incompetent yearlings or in an inadequate number of dominant males being available to fertilize receptive females.

The consequences of these changes are generally unknown. However, red deer hinds mated by young, incompetent males are more likely to be damaged physically than those mated by dominant males. Ginsberg and Milner-Gulland have modeled the effects of female-biased sex ratios on the ability of male ungulates to inseminate females and suggest that it can be a problem at extreme levels (*e.g.*, 12 females to 1 male).

DECIDING WHERE TO LIVE AND HUNT

The distribution of settlements in relation to prey distribution and density has been modeled by Horn and Wilmsen. Their models predict that if resources are predictably and evenly dispersed, then the more efficient pattern of settlement is regular dispersion of small forager social units. In contrast, if the available resources are clumped together and move unpredictably throughout a large range, then the optimal strategy is aggregation of the foraging population at the center of that range.

The ability of humans to directly exchange information about the location of prey resources enhances the value of central-place aggregation. In areas where prey are predictably and evenly dispersed, the concentration of people into fewer larger settlements is likely to reduce harvesting efficiency and result in the depletion of prey populations close to villages. As prey abundances decrease, the range of prey species harvested will increase. By reducing travel costs, hunters can be encouraged to travel farther afield in search of prey.

Thus, if even harvest of resources is the preferred pattern of utilization, effective management would encourage the proliferation of small settlements or the introduction of more efficient means of transport (cars, snowmobiles, motorized canoes, etc.) to reduce search and travel costs. Alternatively, it might be preferable to concentrate harvest in a few areas, around settlements, leaving other areas as prey refuges, depending on the distribution of prey, prey selection, and harvesting patterns.

Many hunting societies have become concentrated in a few permanent settlements that are larger than the traditional settlement units as a result of economic and in some cases government inducements (*e.g.*, the Cree). Natural environments are patchy at some scale or another, with the result that prey are not distributed evenly across the available habitat and hunters must decide where to hunt and how long to stay in each area.

Deciding where to hunt in a patchy or heterogeneous habitat has been the subject of optimal foraging models using the marginal value theorem. These models predict that hunters should remain in a patch as long as the expected returns from the next unit of foraging time in the patch are higher than the expected returns from searching for and exploiting other patches.

In practice, this means that it is not normally worthwhile for hunters to harvest a patch completely before moving on to a new area, with the result that some animals are left (a potential breeding population) in each patch. For example, Cree only shoot the most readily available muskrats before moving on to the next population concentration. Similarly, although beaver lodges may have three to five underwater openings, snares are set at only two or three, suggesting that the return is greater on two snares at the next beaver lodge than on additional snares at the same lodge.

Differences in hunting success in different habitats can dramatically influence hunters' decisions about where to hunt and the prey species caught. Snaring and netting, for example, tend to take place in wooded habitats, where animals often move along defined trails where traps can be set. As a result, in the Serengeti, Tanzania, ungulates living in woodland are far more vulnerable to overharvesting than those living in mixed habitat or open areas.

Deciding how far to travel in search of prey is another problem faced by hunters and one that is strongly dependent on the available information about the distribution of prey resources and the costs of travel.

Central-place foraging models predict that, as prey resources become depleted close to a settlement, the optimal forager will travel farther whenever the cost is offset by the harvesting benefits of more distant hunting sites. The lower the travel costs, the greater the evenness in resource use around a settlement.

Many traditional hunting societies were reported to maintain unhunted areas as sources of future prey or to rotate harvesting areas to provide opportunities for prey populations to recover.

For example, the Mbuti divided their hunting territories so as to maintain a central "no mass land" which acted as a wildlife sanctuary where no hunting took place while the Montagnais-Naskapi randomized hunting excursions and avoided repeated hunts in one area.

Differences in harvesting patterns can substantially affect the impact that different people have on prey populations. For example, while Latin American *campesinos* tend to be sedentary, indigenous people harvest over a wider area, are less sedentary, and harvest on a rotational system, allowing populations time to recover.

Thus, for a variety of reasons, harvesting is often patchily distributed, being more intense close to villages and markets, and resulting in source and sink areas. Many hunted species may be able to tolerate high levels of exploitation as a result of such heterogeneous spatial distribution of the hunting pressure and this may therefore favour prey persistence.

LIMITING HARVESTING TO SPECIFIC TIMES

In general, subsistence harvesting tends to take place throughout the year, although it may be more intensive at some times than others (for example, when other food is in short supply or when less time is required for tending crops. However, harvesting specific species may only be profitable at particular times of year.

For example, although Ache men always pursue armadillos when they are encountered above ground, they only dig them from their burrows, a very labour-intensive process, during the late warm-wet and early dry-cold season when their prey are fat.

The effect of harvesting on prey species is likely to vary through the year. Compared with species that breed over longer time periods, species that have a very peaked birth or mating season may be more severely affected by harvesting than would be expected from the number of animals removed. Harvesting during the rutting season can result in the disruption of territorial structure, increased male-male conflict, and, as a result, reduced rates of conception.

Females not inseminated during their first cycle may either not conceive in that year or continue cycling, which could lead to a decrease in birth synchrony. For example, hunting of saiga *Saiga tatarica* during the mating season resulted in a reduction in conception rates from 85 per cent in 1 year olds and 96 per cent in adults to 55 per cent and 86 per cent, respectively.

Birthing is often carefully timed to fit in with seasonal change, for example, the onset of spring or the rainy season, or is highly synchronized to reduce predation. Consequently, delays due to disruption of normal mating patterns can result in increased juvenile mortality and a consequent decline in population size.

For example, wildebeest calves born outside the main birthing peak experience higher mortality rates as a result of predation. In red deer, a delay in breeding for a single cycle of 18 days can result in a 36 per cent decline in a female's reproductive success as a result of increased calf mortality and a mother's reduced fertility the following year.

The timing of harvesting with respect to the main period of adult mortality can also influence the effect of harvesting on a prey population. For example, in the Serengeti, Tanzania, the main period of poaching is early in the dry season, preceding the period when intraspecific competition for scarce high-quality food would lead to increased wildebeest mortality. Illegal hunting therefore reduces the intensity of intraspecific competition.

DEFENCE OF HUNTING TERRITORIES

Some hunter-gatherer societies defend hunting territories while others do not. Behavioural ecological models predict that territoriality should be found only where resources are economically defendable—that is, when the benefits of defence outweigh the costs. Moderately dense, predictable and evenly distributed resources create conditions that lead to the evolution of territoriality.

Once territories are established, they create conditions more favourable to long-term conservation of resources because the forager is assured of the delayed benefit of short-term restraint. Consequently, conservation of overdepleted prey species may be more likely when hunting territories are defended.

SELECTING PREY BASED ON RELIGIOUS, TRIBAL AND PERSONAL PREFERENCES

Religion not only influences whether wild animals are considered suitable food, it also dictates the type of species that can be harvested. For example, many Pemón indians are Adventists, which permits the hunting of only birds and deer among the higher vertebrates. Bushbabies are avoided in Bioko as being evil and a variety of taboos may prevent the harvesting of particular species, such as chimpanzees in Korup. Individual preferences have a considerable effect on prey-species composition.

For example, the Mayas avoid tapirs because the meat tastes unpleasant. Some food taboos may have their origins in long-term information transfer about the low profitability of rarely encountered prey or in the fact that some species are more vulnerable to overharvesting than others.

COMMERCIALIZATION

Commercialization not only results in the intensification of hunting efforts but also in the concentration on medium-sized species, which can easily be transported to and sold at market; for example, 80 per cent of antelopes but only 10 per cent of rodents go to market in Bioko. The extent to which meat is shipped to market rather than being sold locally or consumed by the family depends on the buying power of the market.

USE OF MORE INTENSIVE HUNTING METHODS

Although the use of shotguns is often mentioned as one of the major factors contributing to overhunting the use of modern weapons does not necessarily result in increased offtake levels but it may increase success rates of hunters. However, some taxa may be more vulnerable than others to overharvesting by shooting.

Primates tend to be very visible, they are highly mobile and difficult to catch with traditional weapons, but they are within range of modern shotguns. Winterhalder reports how access to nets, steel traps, and firearms substantially raised the speed and reliability with which the Cree could catch fish and game once they were encountered, lowering their handling times.

As a result, a larger range of species could be taken, as prey that had previously been considered too small or too difficult to capture were now suitable foraging targets. Subsequently, access to boats and snowmobiles decreased the search time associated with foraging and reversed the earlier trend towards increasing diet breadth. As search time decreased, foragers grew more selective in their choice of prey.

SUBSISTENCE HARVESTING AND BEHAVIOURAL ECOLOGY

While in theory the rate at which a prey species can be harvested sustainably depends on its intrinsic rate of increase, the level at which it is harvested is strongly dependent on the behaviour of both prey and predator. Understanding the factors that influence the susceptibility of different prey species to harvesting and the extent and organization of human hunting behaviour is therefore essential for the development of an effective harvesting management Programme.

Behavioural ecology has emphasized the different roles that individuals play in a population and how reproductive success varies according to age, sex, and individual status. Consequently, behavioural ecology can also help determine how the impact of harvesting depends on both the number and type of individuals removed.

At the moment, the management of subsistence harvesting is usually restricted to broad recommendations such as setting up nonextractive areas, completely banning harvesting, or preferentially selecting males. Information from behavioural ecologists will enable wildlife managers to refine offtake models, make more accurate predictions as to the impact of harvesting on prey populations, and refine management recommendations with respect to individual species.

The problem, of course, is that behavioural data are often difficult to obtain, particularly in tropical forests where much subsistence harvesting takes place. Tropical forest species are often solitary and shy, making behavioural research extremely time consuming and often requiring the use of expensive radio-tracking equipment to maintain contact with the animals.

In addition, difficulties are exacerbated in tropical situations where resources are lacking and species are often close to extinction before action is taken. Immediate decisions are often required without recourse to information from detailed field studies.

However, in some cases, detailed knowledge of one species may enable harvesting recommendations to be generalized to other related species, based on only limited behavioural information.

Realistically, it is in the understanding of human hunting behaviour that behavioural ecology probably has most to offer conservation managers in the near future. Optimal foraging models can help identify decision rules that hunter-gatherers appear to follow in their efficient use of resources and can suggest specific and effective management practices. The success of management Programmes will depend on their acceptance by hunters, and it is easier to persuade hunters to accept management recommendations that minimize the disruption of their optimal hunting behaviour.

Thus, for example, banning the hunting of a species that is not highly ranked is unlikely to cause great contention, but the conservation of a more favoured species may require a more sensitive approach.

In some cases, optimal foraging models have identified circumstances in which game conservation is consistent with maximally effective exploitation. Consequently, little or no intervention is required in the interests of conservation. For example, in patchy habitats, the efficient forager leaves some animals, a potential breeding population, in each patch. An optimal forager from a central place does not necessarily devastate resources close to home and ignore those farther away. Alternatively, diet-breadth models suggest that certain highly ranked prey species will be pursued whenever encountered, and therefore conservation management may be necessary for these. In addition, optimal foraging models can often help predict how a change in hunting behaviour—for example, the use of more effective weapons or more effective transport—is likely to influence the pattern and intensity of hunting.

They may also help explain how demographic processes such as increases or decreases in population density are likely to influence animal communities and human foraging patterns and under what conditions hunters might deplete particular prey resources to extinction.

RECOMMENDATIONS

Although it is clear that behavioural ecology can contribute much to understanding and predicting the effects of harvesting on prey species, it is less clear to what extent wildlife managers will make use of this information. Controlling the extent and nature of subsistence harvesting is notoriously difficult, particularly in the tropical forests of the developing world where many threatened species exist and infrastructure is limited.

Not only will it be difficult to enforce recommendations but, in addition, we are often dealing with situations where people are hunting not to improve their standard of living but for the food necessary to stay alive.

Consequently, we need realistic recommendations that provide sensible options for local people and that are likely to be adopted. Only if behavioural ecology can help conservation biologists make such recommendations will it have contributed anything to the conservation of threatened species. Wildlife harvesting is an important subsistence activity in many parts of the world. Although subsistence hunting poses a major threat to some prey species, the continued freedom to hunt is essential for maintaining the traditional lifestyles of many people as a source of food and income and to control crop predators. It is also one of the few ways in which local communities can derive benefits from wildlife, offsetting some of the direct and indirect costs of forest conservation.

Although in theory the rate at which a prey species can be harvested sustainably depends on its intrinsic rate of increase, the level at which it is actually harvested strongly depends on the behaviour of both prey and predator. The response of a species to harvesting is influenced by behaviour, the number and type of individuals removed, and the role that these individuals play in the population. Understanding the factors influencing human hunting behaviour may enable managers to understand what determines prey selectivity and how variations in habitat, prey availability, and hunting methods influence the impact of hunting on prey populations.

Behavioural factors that may influence inter- and intraspecific vulnerability to harvesting include social behaviour, vigilance and flight, activity patterns, use of trails, ranging behaviour, habitat choice, and attraction to baits. Many species are reported to respond to harvesting by changes in behaviour such as increased shyness, which reduces their susceptibility to harvesting.

However, these changes may have detrimental effects on reproductive rates as a result of individuals trading off the costs of reduced foraging efficiency or parental care against the benefits of reduced predation risk, the indirect effects of human predation.

The determinants of human hunting behaviour have been explored and the consequences for prey populations outlined. Optimal foraging models have been used to explain the extent of subsistence harvesting versus plant gathering and farming and the type of prey that should be harvested, including its age and sex. Predictions include the fact that large prey are generally more profitable than small prey, and low-ranked resources should drop out of the diet when search costs decrease and hence overall return rate increases. The body size of harvested prey is important in determining susceptibility to overharvesting because larger species tend to have lower reproductive rates and therefore are more vulnerable. Behavioural ecological data suggest that the consequences of harvesting particular age and sex classes may not be as clear as originally thought.

Optimal foraging models have also been used to explain the spatial pattern of hunting, how long to stay in a particular resource patch, how far to travel in search of prey, and the role of travel costs in these decisions. Such decisions are important because many hunted species may be able to tolerate high levels of exploitation only as a result of heterogeneous spatial distribution of the hunting pressure. Other aspects of human hunting include the timing of harvesting, the defence of hunting territories, commercialization of hunting, the use of intensive hunting methods, and the role of religious, tribal, and personal preferences in prey choice. Optimal foraging models have also been combined with population biology models to simulate the dynamic relationships between hunter-gatherers and their prey resources.

Bibliography

Allen, C.: *Cognitive Ethology and the Intentionality of Animal Behavior*, Bombay: Manaktalas, 2006.

Asquith, T. J.: *The Inevitability and Utility of Anthropomorphism in Description of Primate Behaviour*, Cambridge: Cambridge University Press, 2004.

Bardi, M.: *Maternal Care and Development of Stress Response in Baboons*, Hyderabad: Economic and Political Weekly Press, 2005.

Burghardt, G.M.: *Animal Awareness: Perceptions and Historical Perspective*, Cambridge: Cambridge University Press, 2003.

Capitanio, J. P.: *Cognitive Style: Problem Solving by Rhesus Macaques*, New Delhi: Concept Publishing Company, 2003.

Case, V.: *Animal Behavior Class Presentation*, Delhi: Motilal Banarsidas Press, 2000.

Dennett, D. C.: *Intentional Systems in Cognitive Ethology: The Panglossian Paradigm*, Canada: Cambridge University Press, 2004.

Griffin, D. R.: *Animal Mind-human Mind*, Berlin: Dahlem Konferenzen, 2001.

Herzog, J. D.: *The Socialization of Juveniles in Primate and Foraging Societies: Implications for Contemporary Education*, Kolkotta: Kolkotta University, 2002.

Heyes, C.: *Contrasting Approaches to the Legitimation of Intentional Language Within Comparative Psychology*, New York: American Geographical Society, 2001.

Latham, N. R. M.: *Maternal Deprivation and the Development of Stereotypic Behaviour*, Maharashtra: Pune University, 2000.

Maestripieri, D.: *Effects of Early Experience on Female Behavioral and Reproductive Development in Rhesus Macaques*, Hyderabad: Economic and Political Weekly Press, 2005.

Pulliainen, E.: *A Contribution to the Study of the Social Behavior of the Wolf*, Karnataka: Regional Resources Centre for Folk Performing Arts, 2002.

Rogers, J. S.: *Genetic Influences on Behavioral Inhibition and Anxiety in Juvenile Rhesus Macaques*, Hyderabad: Economic and Political Weekly Press, 2005.

Scott, J. P.: *The Process of Primary Socialization in Canine and Human Infants*, New York: The Macmillan Company, 2001.

Stevens, H. E.: *Risk and Resilience: Early Manipulation of Macaque Social Experience and Persistent Behavioral and Neurophysiological Outcomes*, Cambridge MA: Perseus Publishing, 2006.

Woolpy, J. H.: *Wolf Socialization: A Study of Temperament in a Wild Social Species*, Maharashtra: Pune University, 2002.

Index

H

I

J

L

M

N

P

R

S

T